Productivity Through Work Innovations

Work in America Institute's National Policy Studies

With the publication in fall 1982 of Work in America Institute's newest policy study, *Productivity through Work Innovations*, the Institute will have completed its fourth major study since January 1978. Previous studies are:

Job Strategies for Urban Youth: Sixteen Pilot Programs for Action
The Future of Older Workers in America
New Work Schedules for a Changing Society

A companion volume to *Productivity through Work Innovations* is *The Innovative Organization: Productivity Programs in Action*, edited by Robert Zager and Michael P. Rosow, the first volume in the new Pergamon Press/Work in America Institute Series.
All of the above books are available from Pergamon Press, Inc.

About Work in America Institute

Work in America Institute, Inc., a nonprofit, nonpartisan organization, was founded in 1975 to advance productivity and the quality of working life. It has a broad base of support from business, unions, government agencies, universities, and foundations, as reflected in its board of directors, academic advisory committee, and roster of sponsoring organizations.
Through a series of policy studies, education and training programs, an extensive information resource, and a broad range of publications, the Institute has focused on the greater effectiveness of organizations through the improved use of human resources.

Productivity Through Work Innovations

**A Work in America Institute
Policy Study**

Directed by Jerome M. Rosow, President
and Robert Zager, Vice-President
for Policy Studies and Technical Assistance

Pergamon Press
New York • Oxford • Toronto • Sydney • Frankfurt • Paris

Pergamon Press Offices:

U.S.A. Pergamon Press Inc., Maxwell House, Fairview Park,
Elmsford, New York 10523, U.S.A.

U.K. Pergamon Press Ltd., Headington Hill Hall,
Oxford OX3 0BW, England

CANADA Pergamon Press Canada Ltd., Suite 104, 150 Consumers Road,
Willowdale, Ontario M2J 1P9, Canada

AUSTRALIA Pergamon Press (Aust.) Pty. Ltd., P.O. Box 544,
Potts Point, NSW 2011, Australia

FRANCE Pergamon Press SARL, 24 rue des Ecoles,
75240 Paris, Cedex 05, France

FEDERAL REPUBLIC Pergamon Press GmbH, Hammerweg 6
OF GERMANY 6242 Kronberg/Taunus, Federal Republic of Germany

The material for the publications produced as part of this project was
prepared under grants from the Andrew W. Mellon Foundation and the
J. Howard Pew Freedom Trust. Phillip Morris, Inc., and Bethlehem Steel
Corporation also contributed to this project.

Library of Congress Cataloging in Publication Data

Main entry under title:

Productivity through work innovations.

   (A Work in America Institute policy study)
   1.Organizational change. 2. Quality of work life--
United States. 3. Labor productivity--United States.
I. Rosow, Jerome M. II. Zager, Robert. III. Work in
America Institute. IV. Series
HD58.8.P76   1982   653.3'14   82-15009
ISBN 0-08-029545-2

Printed in the United States of America

Contents

v

NATIONAL ADVISORY COMMITTEE
PRODUCTIVITY THROUGH WORK INNOVATIONS

Associate Director of the Policy Study
Robert H. Guest
Professor, Amos Tuck School of Business
  Administration
Dartmouth College

*Ruth Antoniades
Associate Director of Social Services
Amalgamated Clothing & Textile Workers Union

*Al Bilik
Deputy to the Director
District Council 37 (New York City)
AFSCME

*Irving Bluestone
University Professor of Labor Studies
Wayne State University

Benjamin C. Boylston
Assistant Vice-President, Industrial Relations
Bethlehem Steel Corporation

*Keith Brooke
Director, North American Work Life Activities
General Motors Corporation

*Sam Camens
Assistant to the President
United Steelworkers of America

William M. E. Clarkson
Chairman and Chief Executive Officer
Graphic Controls Corporation

Donald Ephlin
Vice President;
Director, National Ford Department
United Automobile Workers of America

*Forrest Fryer
Vice-President
Citibank

Nancy L. George
Assistant Postmaster General, Employee Relations
  Department
United States Postal Service

*Malcolm Gillette
Director, Human Resources
American Telephone & Telegraph Company

Victor Gotbaum
Executive Director
District Council 37 (New York City)
AFSCME

*James E. Hamerstone
Director of Employee Relations
TRW Bearings Division
TRW, Inc.

Wayne L. Horvitz
Horvitz and Schmertz
Washington, D.C.

Gene Kofke
Director, Human Resources
American Telephone & Telegraph Company

Delmar L. Landen
President
Delmar L. Landen and Associates
(Former Director, Organizational Research and
  Development, General Motors Corporation)
Detroit, Michigan

Elliot Liebow
Chief, Center for Work and Mental Health
National Institute of Mental Health

Stanley N. Lundine
U.S. House of Representatives

Lloyd McBride
President
United Steelworkers of America

Joyce D. Miller
Vice-President, and Director of Social Service
Amalgamated Clothing and Textile Workers Union

*Herbert S. Millstein
Project Director
U.S. General Accounting Office

Daniel Riordan
Director, Human Resources
Guardian Life Insurance Company

John C. Rudy
Vice-President, Management Services
Kearney-National, Inc.

*Linda Sharkey
PDQ Committee
New York State OER/PEF Joint Labor-
  Management Committee

Tom Smith
Diagnostic Operations Manager
Abbott Laboratories

Harold J. Tragash
Manager, Organization Effectiveness and Research
Xerox Corporation

Eric L. Trist
Professor Emeritus
The Wharton School
University of Pennsylvania,

Brian Usilaner
Associate Director for Productivity
U.S. General Accounting Office

*John Zalusky
Economist, Research Department
AFL-CIO

*Associate Member

LETTER FROM THE BOARD OF DIRECTORS
WORK IN AMERICA INSTITUTE

In the seven years since Work in America Institute was founded to advance productivity and quality of working life, the work-innovation movement has come of age. What was then little more than a gleam in the eyes of visionaries is today a fact of life. Work innovations *work*; they *are* working in enterprises large and small, all over the United States— predominantly in the for-profit sector, but gradually moving into the others. Although still an art rather than a science—probably work innovations will always be so—they are now well enough understood to allow practical guidelines to be formulated with confidence. *Productivity through Work Innovations* celebrates this coming of age.

What accounts for this unexpectedly rapid development? The strongest motivation has come from changes in the economic environment and from the proven value of the innovations themselves. But while the economic problems have been all too evident, the work innovations and their significance have remained relatively obscure. The Institute has made it a point since 1975, therefore, to discover and publicize outstanding programs.

During the 1970s the U.S. economy came up against the harsh realities of the outer world.

First, fuel prices zoomed and price relationships throughout the economy were thrown into turmoil, compelling every major firm to rethink its cost structure. Government efforts to hold domestic fuel prices below the world market helped to sharpen inflation. As productivity-raising machines and systems grew more expensive, employers turned toward work innovations as a less costly alternative.

Next, foreign companies, whose productivity growth had outpaced ours, discovered that large segments of the U.S. marketplace were vulnerable to competition. The strength of the foreign companies was based, in no small measure, on harmonious relations between employers and unions and, particularly in the case of the Japanese, on the thorough involvement of employees in decisions affecting the work itself.

Third, the invasion of our markets here and abroad by automobiles, electronics, optics, and other high-technology products was achieved because foreign companies outperformed U.S. companies in product quality, delivery, marketing, and customer service, areas in which this country's managerial skills had been considered supreme.

Fourth, labor-management cooperation also enabled foreign competitors to introduce advanced technology more rapidly than their U.S. counterparts. This augmented the economies of scale gained through market penetration.

Lastly, the work force changed, not only in age-mix, sex-balance, and educational level, but also in values and expectations, most of all regarding the perceived right to participate in decisions on the job.

The impact of these blows shook many business leaders out of their confidence in the superiority of American methods. Some CEOs were moved to test new alternatives, especially in the use of human resources.

Meanwhile, a few leading companies had already shown the way. Of these, the United Auto Workers-General Motors example was the most persuasive because of the company's size and prominence; the scope, variety, and daring of the work-innovation program; and the fact that the union, noted for militancy and inventiveness, was a full partner in every aspect of the program.

Today, the prime need is to channel the movement away from faddism and fantasies of quick success and toward the path of steady growth and development. The Institute's policy report, *Productivity through Work Innovations*, and its accompanying casebook, *The Innovative Organization: Productivity Programs in Action*, seek to meet this need. They zero in on the dynamics of work innovation and the divergent interests to be reconciled. With 48 specific, practical recommendations, the report guides decision makers step by step through the strategy and tactics of launching, diffusing, and institutionalizing change. The recommendations are underpinned by a wealth of concrete experience, as exemplified by the dozen lively cases in *The Innovative Organization*.

We are pleased to present the report and casebook for the attention of executives in industry, labor unions, government, and the other sectors of the economy. The discussion is pragmatic, timely, and pertinent to the needs of our society, which is earnestly trying to enhance productivity and the quality of working life. It seems most fitting to us as a Board that the Institute, founded in 1975 to promote those twin goals, is now able to provide the insights and down-to-earth guidance that will bring them into the workplace.

Work in America Institute, Inc.
Board of Directors

Ronald H. Brown

Irving Bluestone

H. W. Clarke, Jr.

Emilio G. Collado

Thomas R. Donahue

Clark Kerr

G.G. Michelson

Elliot L. Richardson

Jerome M. Rosow

Daniel Yankelovich

Preface

Productivity through Work Innovations is the report of the fourth national policy study conducted by Work in America Institute. The complete endorsement of the report by our Board of Directors reflects the support of respected representatives of labor, management, and the general public.

The object of the report is not to lay out alternatives or advocate ideal solutions, but to recommend and define practical courses of action based on "the state of the art" and tested against the experience of knowledgeable leaders. It is designed to help decision makers in all types of organizations to act with enlightened self-interest as they consider, plan, implement, diffuse, and institutionalize work innovations. In a sense, this report captures the leading-edge workplace innovations of the past decade and distills the ideas that work into a handbook for change.

A companion volume, *The Innovative Organization: Productivity Programs in Action* by Robert Zager and Michael P. Rosow, presents 12 outstanding case studies of work innovations and provides the foundation of experience on which the report itself rests.

In connection with the policy study, we wish to express our thanks to the following:

—The Andrew W. Mellon Foundation, the J. Howard Pew Freedom Trust, Phillip Morris, Inc., and Bethlehem Steel, for their financial support of the study. The statements made and views expressed in this report, however, are solely the responsibility of Work in America Institute.

—Our National Advisory Committee, whose names have been listed in these pages, for their faithful and vigorous participation.

—Professor Robert H. Guest, for serving as associate director of the study and for preparing drafts of chapters 1, 6, and 7 of the report. (The Institute alone, however, is responsible for the opinions contained herein.)

—Paul E. Barton, for preparing drafts of chapters 3, 4, 5, and 8.
—Professor Irving Bluestone, for his discussion paper on "The Union and the Quality of Worklife Process."
—William Clarkson and James F. Gillespie, for their discussion paper on "The CEO's Impact on Productivity."
—Professor Paul H. Goodman and James W. Dean, Jr., for their discussion paper on "Institutionalization, or Making Labor and Management Change Programs Last."
—James E. Hamerstone, for his discussion paper on "Work Teams."
—Wayne L. Horvitz, for his discussion paper on "Some Concerns about Cooperative Work Innovations and Labor-Management Relations."
—Arnold F. Kanarick, Ph.D., for his discussion paper on "Quality Circles in Honeywell."
—Professor Edward E. Lawler and Addison-Wesley Publishing Company for permission to use as a discussion paper the chapter "Gain Sharing" in Professor Lawler's recently published book, *Pay and Organization Development* (1982).

We particularly acknowledge the outstanding contribution of Robert Zager, who, as Work in America Institute's vice-president for policy studies, determined the parameters of the study, presided over advisory committee meetings with unfailing patience and good humor, and skillfully combined various viewpoints in a report that provides a road map through much of the hitherto unexplored territory of innovative work practices.

We wish also to thank those members of the staff of Work in America Institute who were associated with this project. Without their skill, energy, and devotion, the study could not have been completed:

   —Beatrice Walfish, editorial director, and Frances Harte, managing editor
   —Joan White and Stephanie McDowell, typesetting and production staff
   —Virginia Lentini, who, as assistant to the vice-president for policy studies, managed the logistics of the study
   —Meredythe Nowak, Kathleen McDonald, and Carol Nardi, executive support staff, who provided word-processing assistance

JEROME M. ROSOW
President
Work in America Institute

Summary/Recommendations

Hundreds and hundreds of employers in the United States today, large and small alike, are systematically employing the supervisory and managerial abilities of rank-and-file workers and profiting thereby. They are doing this through quality-of-working-life approaches commonly called "work innovations"—a variety of ways to make jobs more autonomous, more creative, and more demanding of the skills of the people who fill them. They range from the redesign of individual jobs to the formation of consultative and decision-making groups, more commonly known as quality circles, employee involvement teams, labor-management committees, or semiautonomous work teams.

The numbers of employers who are trying work innovations will grow rapidly for the simple reason that the harnessing of workers' brainpower is advantageous to employers, employees, unions, and the general public. Those who interpret these developments as a passing phase related to economic recession are misreading the signals. The economic environment of American industry has changed permanently, most of all for those in the business of mass production and/or high technology, namely the pattern setters. Companies that compete in today's and tomorrow's markets have no choice but to raise sharply the quality standards of their products and services without raising costs. And there is no known way to accomplish these goals except to engage every member of the enterprise, from top to bottom, in this pursuit.

General Motors was persuaded to that effect by the United Automobile Workers (UAW) in 1973, long before the automotive

1

crisis struck. Chrysler and Ford, despite formally assenting to QWL in the same negotiations, did not practice it until several years later. Now, nothing short of economic disaster will cause the giants to renege on their commitments to work innovations. With their example looming so large, the rest of industry has begun to follow suit, or at least to reexamine the alternatives and accept the necessity for change.

Other major forces, related both to the economy and to the nature of a new American work force, are also impelling industry in this direction.

Underlying Forces

A number of economic and social forces have contributed to the increased use of work innovations in recent years. Most of these derive from the persistent decline in productivity growth rates since 1965 and increasing awareness of the need for productivity improvement. Many of them are interrelated and, taken together, have acted as a force for both experimentation and change.

☐ *The economy of the 1980s.* Slow economic growth, high energy costs, and record-breaking capital-investment costs linked to high and unstable interest rates are restraining investment in the private sector. Uncertain monetary and fiscal policies have also added to the necessity for more efficient management *within* the enterprise itself.

☐ *The productivity crunch.* Since 1965, U.S. productivity growth rates have declined; during 1979-81, productivity itself declined. While U.S. productivity in absolute terms is still the highest in the world, other nations are increasing their productivity faster than we are and are rapidly closing the gap. The search for higher productivity has led to an interest in the more effective use of human resources.

☐ *Technological change.* The rate of technological change has accelerated and will affect office, factory, and the executive suite, forcing management to rethink the way people are used in relation to technology.

☐ *Quality of product and service.* The shift from a society of low prices and planned obsolescence to a society of high prices and the demand for durability has created a new production imperative— quality of product and service. Quality can best be attained, employers have found, by including their employees in decisions on how best to reach this goal.

☐ *International competition.* Increased imports are exerting continuing pressure on U.S. industry to improve its position in both domestic and foreign markets by producing the higher-quality products now demanded by consumers.

☐ *Labor costs.* A higher proportion of professional, technical, and skilled personnel in the occupational mix, combined with wage inflation and expanded benefit coverage, has created pressures for better utilization of personnel.

☐ *Job security.* Employees are becoming aware that the continuity of their work lives is dependent on the continuity of the enterprise. Thus they are increasingly interested in participating in management efforts to keep the enterprise alive.

The Work Force of the 1980s

There has been a marked change in the composition and quality of the work force during the last decade—as well as in its attitudes and values. This new-breed work force has exerted a particularly strong force for change in the workplace. The following elements most clearly define the character of the new work force:

☐ *Changed demographics.* The work force will show a sharp increase in the 25-44 age bracket, a decline in young workers, and a continued increase in the participation of women. The median age will increase steadily.

☐ *Role of women.* The combination of education and the economic pressure to work has sharply changed the role of women in the last decade. Women, who now comprise about 42 percent of the labor force, are continuing to move into the workplace faster than men. In the year 1979 alone, one and a half million women entered the labor force.

☐ *Education.* The American work force is the most highly educated in the world and will continue to improve in quality. In the past decade, high school graduates in the work force rose from 32 to 42 percent and college graduates from 10 to 18 percent. Twenty-one million Americans have completed four or more years of college, and college enrollment is at an all-time high of 12 million.

☐ *Attitudes and values.* There has been a marked and persistent shift in the attitudes of younger workers, which is now spreading to the work force as a whole. The work ethic—a basic belief in the value of work—has not diminished, but many workers feel that there is an incongruity between more flexible modes of life and expression in the outside community and the authoritative, technocratically controlled workplace. They do not oppose the legitimate exercise of authority, but they do resent "authoritarianism," a distinction that is crucial to worker-manager relations and national productivity.

Since American workers live in a free and open political system, they expect conditions in the workplace to be compatible with the society. Their proper expectations include the right to free speech, privacy, dissent, fair and equitable treatment, and due process in work-related activities.

A majority of all Americans today believe that they have a right to take part in decisions affecting their jobs.

The Spectrum of Work Innovations

The unifying theme of work innovations is the attempt to design into the jobs of all employees those characteristics which have hitherto been reserved for professional and managerial jobs— variety, relative autonomy, identifiable product, scope for creativity, and judgment. Work innovations discard the "scientific management" concept of one-man/one-motion. There is no one best form of innovation; each organization must tailor a solution to its own unique requirements.

Many approaches have been tried, each presenting advantages and disadvantages.

☐ *Individual job designs* include *job rotation* (increasing the number of not-necessarily-related tasks), *job enlargement* (lengthening the time cycle by adding to the number of different steps a worker performs in sequence), *job enrichment* (increasing responsibility for decision making).

☐ *Group job designs* are consultative and/or decision making.
—*Consultative* groups study and recommend solutions to problems of importance to a work unit or an establishment, while management makes the decisions. Examples are quality circles, employee-involvement teams, problem-solving teams, labor-management committees.
—*Decision-making* groups are responsible for putting into effect the solutions they have devised for problems within their jurisdiction. Examples are semiautonomous, or relatively autonomous, work teams.

☐ *Gain-sharing plans*, such as Scanlon, Rucker,® and Improshare,® foster creativity and teamwork by linking monetary rewards to increases in the productivity of the entire establishment.

☐ In the *sociotechnical* approach, the design of jobs is integrated with the design of technology and organization.

This report focuses on consultative and decision-making

groups, which appear to have the greatest potential for increasing productivity through work innovation.

In order to survive in the real world, new job designs must adapt to changing conditions. Successful adaptations evolve in a manner that is cooperative, open, and inventive, and not by fiat. The trend of this evolution is toward:

☐ Joint optimization of technology and organization
☐ The worker as complementary to the machine
☐ The worker as a resource to be developed
☐ Optimum task grouping, with multiple broad skills
☐ Employee self-control
☐ Flat organization chart
☐ Participative style
☐ Collaboration
☐ Commitment
☐ Innovation
☐ Taking into account employees' and society's purposes as well as the organization's

The Players and the Stakes

Although work innovations are usually looked at in terms of particular job designs, the real name of the game is the changing of organizational cultures to make work life both more productive and more compatible with a democratic society. In that game the interests of all the players must be fairly dealt with at the same time.

☐ The *general public* looks to productivity improvement as the source of rising standards of living, and to work innovations as a means of achieving them peacefully. But government cannot legislate organizational change. It can only be supportive of organizations as they take ground-level decisions to change themselves.

☐ The *sine qua non* for every work innovation is the *employer's* decision to undertake it. The decision should be based on careful assessment of pros and cons. Work innovations offer an opportunity for substantially improving organizational performance—especially the quality of product or service—with relatively little capital cost, but they cannot substitute for managerial skills. They also bring "soft" improvements (in absenteeism, turnover, grievances, labor-management relations), which are, in fact, hard and long-lasting economic gains. The main costs are time, personal attention, and, most of all, the strain of learning to share information and power.

☐ *Employees* consistently opt for work innovations once they understand what is involved and are persuaded that the employer's intentions are legitimate. The attractions for them are job security, higher earnings, better relations with supervisors, and opportunities for personal growth. The hazards are the anxiety of accepting broader responsibilities and the possible loss of jobs or earnings due to increased productivity.

☐ A few of the country's largest international *unions* are partners with employers in programs of work innovation. Most internationals remain skeptical but do not inhibit locals from taking part. Their doubts reflect the understandable absence of a ground swell of member demands for work innovations and the suspicion that employers are aiming at a "union-free environment." However, since innovations are very popular with workers *once they have tried them*, unions can gain by taking the lead instead of letting employers "own" the improvement. If they do go in, unions should be full partners from the start.

☐ *Managers and supervisors* have the most to lose in the short term. It is their authority and access to information that will be shared with subordinates, and often they are asked to share while their bosses are not. Unless they perceive that their own quality of working life will eventually benefit from work innovation, they will torpedo it. The key advantage to them is the ability to accomplish much more than before without any real loss of authority.

The Process of Work Innovation

The process of carrying out a quality-of-work-life effort is a carefully planned, step-by-step process that includes three separate phases:

☐ *Launching*
—*Top-level commitment.* The first steps in work innovation are clear commitments by the CEO and top union leaders, who then ensure that all levels of leadership on both sides support the change.
—*Goals of change.* The declared objective should be to tap the creativity of the entire organization for the benefit of all.
—*Grass-roots action.* Programs of work innovation should be introduced first in those areas which are most ready for them and should be congruent with the history and character of the organization. No matter on how small a scale the project begins, the parties must make clear that the ultimate intention is to change the entire organization.

—*Consultative work groups* should be given great latitude in choosing problems and devising solutions, and should receive training suited to their needs. Management's manner of dealing with their recommendations can make or break a program of change.

☐ *Diffusion*

—*Strategy* for diffusion should be drawn up as soon as pilot projects have had a recognizable success and should take into account any problems surfaced by them.

—*Pace* of diffusion is ticklish. It should be as fast as, but no faster than, financial and professional resources can handle.

—*Diffusion within a site* should allow each work unit to adapt the program's design to its own conditions.

—*Diffusion from one site to another* should allow each site to determine its own form and process of work innovation, subject to attaining minimum standards of progress.

—*Recognition of achievements* does wonders in reinforcing work innovations, whether the recipient is a work group or a manager. Both monetary and verbal rewards are effective, singly or in combination.

—*Painstaking attention to communication and education* is essential to the diffusion effort, but there is nothing as persuasive as seeing a successful project with one's own eyes.

☐ *Institutionalization*

—*Plans for institutionalizing change* (i.e., making sure it lasts) should be part of the original strategy for work innovation.

—*Realistic expectations.* Since no organizational change is ever a pure success or a pure failure, the parties should try to agree beforehand about the kinds and degrees of change they hope for.

—*Evidences of institutionalization.* To assess whether a work innovation has taken root, it is necessary to ask: Do employees know what new behaviors they are supposed to perform? Do they perform them? Do they feel that the behaviors are appropriate and have become the norm? Do they feel that the change has been rewarding?

—*A moving stream.* Because the work force and the organization are continually changing, work innovations require training and *re*training, commitment and *re*commitment.

—Dependencies. A work innovation that is excessively dependent on the support of an internal sponsor or an external consultant is likely to collapse as soon as that support is withdrawn.

Issues on the Horizon

As an increasing number of work organizations embark on quality-of-work-life efforts, other problems will come to the fore.

□ *Employment security.* Economic crises in major industries, such as steel, auto, rubber, airlines, and textiles, combined with the record-breaking levels of unemployment, have increased the need of employees for job security. The key to maintaining continuous employment for all employees is for the organization to adopt this objective as a part of its overall strategy.

□ *Gain sharing.* Although the majority of consultative programs contain no explicit gain-sharing features, there is a need to develop a method for the periodic sharing of financial gains directly attributable to work innovation. Ideally, management should have plans for gain sharing ready before employees demand it.

□ *The adoption of work innovations in other sectors of the economy.* For the most part, work innovation has been concentrated in the private, for-profit economy. Only limited efforts have been undertaken in health care, education, and governmental sectors, although there are some notable exceptions. Since these sectors are labor intensive, they provide targets of opportunity for the introduction of work innovations in the 1980s.

A Work in America Institute Study

As a means of accelerating the spread of work innovations, Work in America Institute embarked in 1981 on an eighteen-month study of work innovations, with the principal support of the Andrew W. Mellon Foundation and the J. Howard Pew Freedom Trust. Phillip Morris, Inc., and Bethlehem Steel Corporation also made important contributions.

The study resulted in a major report, *Productivity through Work Innovations*, and a companion casebook entitled *The Innovative Organization: Productivity Programs in Action*, edited by Robert Zager and Michael P. Rosow. The casebook describes the experiences of leading corporations and unions in carrying out work-innovation projects. The policy study report defines and analyzes the policy issues encountered in the process of work innovation and recommends practical ways of accomplishing results.

When this study was first conceived, its objective was to encourage the spread of work innovations. That objective has been supported and accelerated as a result of the expanded use of work innovations in this country, particularly following the arrival of quality circles on our shores. Economic imperatives, combined with actual achievements, have moved these programs from the test-tube stage to the workplace at large.

Thus, work innovations are no longer unknown quantities; they have been tested in the cauldron of experience and they work. The purpose of the report, therefore, is to continue to support positive methods of achieving productivity gains by offering practical guidance to employers and unions as they take up the challenge of involving all employees as full members of the enterprise.

Chapter 1 offers an overview of work innovations in the United States: the underlying forces, the interests and positions of the parties involved, and the spectrum of work innovations that has evolved to meet the need.

Chapters 2 through 5 discuss work innovations from the perspectives of employers, employees, unions, and supervisors and managers.

Chapters 6 through 8 describe the management of work innovations: launching and pilot-testing, diffusion, and institutionalization.

The final chapter looks ahead to issues relating to work innovations that are just below the horizon: job security, gain sharing, and the further spread of innovation. These issues will undoubtedly intensify in the years ahead.

In short, the report is concerned with the multiple ways in which employers utilize the one unlimited resource—the men and women of working America. More specifically, it embraces the proposition that there are boundless opportunities through work innovations to improve productivity at the workplace and to unlock the creative human potential that exists at all levels of our production and service institutions.

Recommendations to employers and unions to assist them in launching and then managing work innovations are incorporated in each of the chapters that follow. They are also listed below for the reader's convenience.

Recommendations

The Employer's Perspective

1 Before entering into work innovations, and from time to time along the way, the employer should try as objectively as possible to assess and reassess the probable gains and risks and the effect on employees. *(Page 42)*

2 An employer considering the advisability of entering into work innovations should not be deterred by the fact that employees are represented by a union. The presence of a union can increase the chances of success, provided that the union (1) is persuaded that the employer does not intend to weaken or dislodge it, and (2) becomes a full partner in the effort. *(Page 43)*

3 The parties should declare publicly, in advance, what they hope to gain from work innovations, and their declarations should be mutually consistent. The employer's goals, in particular, should be candid, realistic, and long term. *(Page 47)*

4 The employer should not ask the work force to increase "productivity," as such, but rather to improve particular aspects of performance that are within its capacity. *(Page 48)*

5 Employers should take seriously the so-called "soft" benefits of work innovations. Improvements in absenteeism, turnover, grievance load, employee motivation, and teamwork produce hard economic gains. These become normal standards of performance rather than temporary or one-time gains. *(Page 50)*

6 Employers should recognize that quality of product or service is increasingly the key to competitiveness, that quality control is increasingly the key to productivity, and that work innovations are indispensable to both. *(Page 53)*

7 Employers should regard the financial costs of work innovations as an investment with the potential for a very high rate of return. They should regard the organizational and personal strains

of adapting to work innovations as difficult—although not insuperable—and worth the effort. *(Page 56)*

The Employees' Perspective

8 Employers should adopt the working hypothesis that most of their employees would, in proper circumstances, respond favorably to the prospect of taking part in work innovations. *(Page 59)*

9 Since cooperation cannot be commanded, and since there is no reliable way of knowing in advance which individual employees are in favor of work innovations, employers should ensure that all participation is genuinely voluntary. *(Page 60)*

10 Employers should adopt work innovations because they bring democratic values into the workplace without jeopardizing the authority management needs in order to operate successfully. Indeed, work innovations strengthen employees' regard for management. *(Page 65)*

11 The CEO should bear in mind that managers and supervisors are also employees who will find it more natural to treat subordinates participatively if they themselves are treated that way. Managers cannot function well under a double standard. *(Page 68)*

The Union's Perspective

12 Since work innovations can serve the interests of workers and unions, as well as contribute to the survival and prosperity of employers, unions should take a public position in favor of work innovations in principle, without dropping their guard with respect to specific applications. *(Page 73)*

13 Unions which are still debating whether to take a public position in favor of work innovations should take note of the potential advantages for union leadership.
   a. Furtherance of the union goal of democratizing the workplace.

b. Direct participation in vital decisions about the enterprise.

c. Strengthened membership due to greater economic security.

d. Greater know-how about the operation of the enterprise.

e. Greater openness between management and union.

f. Better QWL for union officials, because there is a better balance between problem-solving and adversarial activities.

g. Less troublesome negotiation, ratification, and administration of contracts. *(Page 73-74)*

14 The most effective way to counter the abuse of QWL programs by antiunion employers is for unions themselves to take the lead in demanding and implementing QWL programs. *(Page 76)*

15 Union leaders should study and engage in the QWL process while remaining firmly committed to their basic bargaining role. Tough collective bargaining on issues that divide the parties is entirely compatible with day-to-day cooperation on shared interests. *(Page 79)*

16 Unions should be open to the QWL process since it will not alter either their structure or their constitution. *(Page 81)*

17 Unions should maintain a clear line between QWL and collective bargaining, although certain questions arising out of QWL may require settlement through negotiations. Nevertheless, those union officials who are responsible for collective bargaining should also take part in the QWL process. *(Page 83)*

18 In order to establish the legitimacy of work innovations, the union should insist at the outset on being an equal partner with management in all matters pertaining to the project. *(Page 84)*

19 When local union officials believe that management would like to begin a joint program of work innovations, the union should take practical steps to further the aim. For example:

a. Schedule a joint meeting to "break the ice" and establish a climate of trust.

b. Invite union and management representatives experienced in running QWL programs to attend a joint meeting and make presentations. *(Page 85)*

20 When the local union wants to inaugurate a QWL program, but the organization shows no interest, the union should take the initiative in bringing it about. *(Page 86)*

The Perspective of Supervisors and Managers

21 The employer's strategy for work innovation should take account of the impact on supervisors by providing for (1) redefinition of the supervisor's job, (2) reorientation and retraining to meet the new job definition, (3) employment security, and (4) relief from pressures that might compel the supervisor to fall back on traditional methods. *(Page 89)*

22 The employer should redefine the supervisor's job so as to (1) reduce "policing" and "expediting" duties, (2) increase planning, coordinating, and advisory duties, and (3) make it a satisfying whole job. Then the employer should ensure that everyone concerned understands the new definition. *(Page 91)*

23 Since the number of supervisory positions may be reduced, and some supervisors may be capable of broader responsibilities, the employer should assure all current supervisors that they will either receive alternative positions of suitable status and income within the firm, or be assisted until they find suitable employment elsewhere. *(Page 93)*

24 The CEO should commit personal time and energy not only to launching the work innovation but to ensuring that it becomes a permanent way of life for the organization. *(Page 94)*

Preparing, Launching, and Pilot Testing Work Innovations

25 Before an organization sets out on a course of work innovation, the CEO should announce that he or she is committed to that change. This commitment should be for the long term and should include the needed resources. *(Page 99)*

26 Once having declared his or her commitment, the CEO should take the responsibility for ensuring that all levels of management support the proposed change. *(Page 100)*

27 Before undertaking action at the grass roots, the CEO should assess the degree of readiness for change in the various parts of the organization and should, if feasible, use the findings as a basis for initial change activities. The employer should also be careful to choose a form of work innovation that is compatible with the organization's true nature, because a mechanism that requires sharing more information or more decision-making power than management can live with is doomed to failure. *(Page 104)*

28 As soon as the union decides to take part in the work-innovation process, the parties should issue a joint statement which outlines (1) their objectives, and (2) specific safeguards for employees. *(Page 107)*

29 The parties should make sure that everyone concerned understands from the beginning that, even though the process starts in one or a few select areas, the entire organization will ultimately be involved. *(Page 108)*

30 The parties should launch grass-roots action in a small number of work units, using them as pilot projects in which procedures and problems can be sorted out before expansion begins. Units should be selected in which the chances of success are high and where strong leaders are in place. *(Page 111)*

31 Management should deal with work-group recommendations in a timely, respectful, and open-minded fashion. When modification or rejection is called for, management should state its reasons clearly and offer to help if the group wishes. *(Page 114)*

Growth and Diffusion of Work Innovations

32 As soon as the pilot programs have been recognized as successful, the parties at the local level should draw up a strategy for diffusion to the rest of the site, based on the early experiences and tailored to the unique conditions of the organization. *(Page 119)*

33 In diffusing work innovations within a location, the parties should direct their efforts toward work units whose supervisors and employees are most ready and willing to take part. However, they should make clear from the start that all work units at the location will eventually have the opportunity to participate. *(Page 121)*

34 The parties should use all available formal and informal means to acquaint nonparticipant employees with the activities of the groups and should encourage the groups to involve other interested employees in the groups' activities. *(Page 124)*

35 The parties should ensure that the accomplishments of the problem-solving groups receive public recognition from management. *(Page 126)*

36 The parties at corporate level should develop a strategy for diffusing work innovations throughout the enterprise. The strategy should allow each location great latitude in determining its own form, process, and pace of change, provided that reasonable minimum standards of progress are met. It should also provide maximum opportunity for managers to learn about work innovations in all parts of the organization, through site visits, conferences, educational programs, networks, and the like. *(Page 129)*

37 As part of the diffusion strategy, top management should conspicuously reward managers who achieve outstanding performance through work innovations. *(Page 130)*

Institutionalizing Change

38 The employer (and union) should include plans for institutionalization as an integral part of the strategy for introducing work innovations. *(Page 132)*

39 The parties should measure results on the basis of what they really want to achieve, track qualitative results as well as quantitative, and avoid the temptation to overmeasure. *(Page 134)*

40 When introducing work innovations at a traditionally managed site, the parties should strive for congruence, splicing the change with the existing structure, rather than making a superficial graft that won't last. *(Page 136)*

41 The parties should ensure that participants are provided with the requisite minimum of training in new work behaviors before the work innovation is adopted. They should also create mechanisms to ensure that retraining is provided from time to time and that subsequent participants receive such training as well. *(Page 137)*

42   Top leadership of the employing organization (and the union) should demonstrate its commitment to the innovation, publicly and tangibly, in order to encourage voluntary participation and create staying power. *(Page 138)*

43   The parties should include in the strategy of work innovation a system of rewards, both intrinsic and extrinsic, linked to actual performance of desired new behaviors and equitably distributed. However, actual payments of extrinsic rewards should never be made in advance of the achievements on which they are contingent. *(Page 140)*

44   The parties should include in the original strategy mechanisms for creating a flow of information from the program and making adjustments as they are needed. Employee commitment and involvement are sustained by a free and steady flow of information. *(Page 142)*

45   The parties should use consultants when necessary but not let the program become too dependent on them. The consultant's highest contribution is to help create an internal capability, rather than become a substitute for it. Consultants should adapt to the organization rather than require the organization to be re-fashioned in the consultant's mold. *(Page 143)*

46   The parties should ensure that a work innovation does not become permanently identified with its sponsors. However, in the early days of a project, when the role of the sponsor may still be vital, the parties should transfer or move the sponsor only if they have a replacement who is willing and able to carry the project forward. *(Page 144)*

Issues on the Horizon

47   Employers should strengthen employment security in order to create the basis for employee confidence in, and cooperation with, work innovations. Their policies should balance the need for flexibility with the need for security, recognizing the investment in human capital represented by training, experience, and established relationships. Where there is a union, these trade-offs should be collectively bargained. *(Page 151)*

48 Employers engaged in work innovations should be sensitive to the long-term gain-sharing issues which are bound to arise. In cost-reduction sharing plans (with or without unions), a formula is defined and agreed on beforehand. In other programs, gain sharing should remain a secondary goal, but as significant savings and improvements are generated, the employer should agree with the employees (through their union, if there is one) on a method for periodic sharing of financial gains directly attributable to the work innovations. *(Page 153)*

1.
Work Innovations: State of the Art

One of the most hopeful developments in response to declining productivity growth in the United States has been the increasing ability of American industry to innovate in the workplace. This movement represents lasting change in the application and use of human resources in the workplace and in the relationship between unions and management. Change has been accomplished through individual job redesign and through the various forms of consultative and decision-making groups which have mushroomed in American enterprises during the last decade: quality circles, employee involvement teams, labor-management committees, semiautonomous work teams.

Work innovations have attracted the interest of employers because they have proved to be an effective way to increase productivity. In addition to being useful in their own right, they have a synergetic effect on other means of raising productivity, such as new technology, new systems, and new product mixes. Moreover, the capital costs of improvement through work innovations are insignificant by comparison with other means.

It was not always thus. Traditionally, productivity growth has been viewed primarily in classic business terms—better use of capital, more effective marketing strategies, applications of advanced technology, and more efficient management techniques. These techniques were based on the principles of "scientific management" proposed by Frederick Taylor some 80 years ago. Taylor believed that manual operations could be reduced to a set of reliable motions by which the worker, the tool or machine, and the materials could interact in "the one best way." In such a system workers were not allowed discretion of any kind; in fact, to permit decision making would have deflected workers from the established pattern. Planning, directing, coordinating, modifying, measuring, and evaluating were reserved for supervisors and managers.

Although Taylorism dominated mass production technology for the first part of the century, there were always a few employers who were interested in utilizing the ideas, experience, inventiveness, and initiative of workers. Several major unions, even before World War II, joined with employers in experiments of this type. After the war, the number of experiments increased and the anti-Taylor approach came to be called "participation," a word that included so many different types of consultation and decision making—and often only the appearance of participation—that its usage became somewhat less than precise. A newer and more exact term, "work innovations," is now being used to describe programs that provide thinking roles for workers.

In the late 1960s and 1970s, many more employers came to realize that the productivity of the work force was a substantial factor and that traditional methods for stimulating productivity were no longer good enough. Some began to unlock the creative and productive potential of their employees through programs which required a whole new set of assumptions as to what motivates people and what constitutes organizational effectiveness. These innovations required the involvement of all members of an organization in decisions affecting their work, a principle enshrined in management theory but usually honored in the breach.

During the 1970s many executives considered this more egalitarian approach a threat to their status and authority. Acquiescing in it was perceived as a dispersion of power to the lower levels—an act that went against the grain of the traditional manager. Yet General Motors, Honeywell, and dozens of other major corporations applied the principle with great success, and their example encouraged many more to meet the productivity challenge of the 1980s. It seems clear today that a sea change in American industrial practice is underway.

Two sets of forces will accelerate the movement in the coming decade. First, in a period of expensive capital, the mounting competitive demands for higher productivity and higher quality of product and service will make it imperative to use the work force more effectively. Second, changes in the composition and attitudes of the work force will make innovations increasingly attractive to employers, managers, employees, and unions. The underlying forces, the interests and positions of the parties involved, and the spectrum of work innovations that has evolved to meet the need are examined in this chapter.

WHY MORE EFFECTIVE MANAGEMENT OF PEOPLE IS NEEDED

The Economy of the 1980s—The Productivity Crunch

The economy of the 1980s may well be dominated by a series of interlocking crises already highlighted by the 1981-82 recession. Ultimately ways must be found to reverse the slump in American productivity; for it is productivity which sustains the economic health of the nation. While the productivity of the U.S. economy in absolute terms still leads the world, other nations are increasing productivity faster than we and are rapidly closing the gap.

Productivity gains in the 1980s will depend in large part on the availability of capital, rates of return on capital investment, new energy sources, applications of new technology, and, most important for purposes of this report, the productive quality of the work force. Already companies are turning toward more effective use of the work force as a supplement to, or substitute for, capital investment. For example, early in 1980 the chief executive of one of America's largest manufacturing corporations asked permission to send a delegation of senior managers to Work in America Institute for advice on how to improve productivity. When the Institute offered to discuss problems of technological change, he replied: "No, the corporation can't afford major technological change or capital investment; it has to raise productivity through its employees."

One of the new lessons for managers is that under today's conditions "productivity" and "quality of product or service" are not different things but merely different ways of looking at the same thing. In many cases, quality of product or service is the key to survival, and, as the Japanese have made painfully clear, work innovations play an essential part in the improvement of quality.

Present estimates suggest that the rate of productivity growth is unlikely to return to that of the two decades preceding 1966. Annual productivity growth of total private business, which averaged 3.3 percent from 1947 to 1965, slowed to 2.2 percent from 1965 to 1973, fell to minus 0.3 percent in 1979 (the steepest plunge since the Bureau of Labor Statistics began compiling productivity statistics in 1947),[1] and continued at negative levels in 1980. The

1981 increase of 1.1 percent was the first improvement.[2] Optimists predict that productivity will rise again, but probably not above 2.5 percent by 1985. It is worth noting that even during the period of our fastest productivity growth—from 1947 to 1955—the U.S. rate was far below that achieved by Japan, Sweden, West Germany, France, and others during the 1970s.[3]

Observers have attributed the slowdown to cyclical factors, to shifts in the types of industries in the private sector and their level of output, and to the entrance into the labor force of large numbers of relatively unskilled youth and women employed in less productive, service-type occupations. No one has made a convincing case that the decline is due to lessened individual work effort. The United States will not decline economically, if decline it does, because of the labor force's unwillingness to work reasonably hard.

Labor has been viewed generally as a cost in the process of production, a cost which must be managed. Except as a cost factor, *labor is rarely seen as a critical dynamic factor in the productive outcome*. This negative attitude narrows the options for action with and through people.

The human factor in productivity is subtle and usually under-estimated, if not ignored. Economist Edward Denison estimates that during the years 1948 to 1973, the human factor—a combination of quantity and quality of labor—was responsible for nearly 29 percent of the productivity growth rate.[4] People account for 50 percent or more of controllable costs; in labor-intensive service and government, they account for 70 to 85 percent of all costs.[5] Computerization and robotization will make for leaner organizations in terms of the number of employees needed; but, paradoxically, the leaner the organization, the heavier its reliance on each individual employee.

The next decade will see an explosion of new technologies, with enormous potential for productivity growth. Although the rate of introduction may be slowed by high interest rates and the high cost of capital relative to return on investment, lower tax rates and technological competition should provide incentives. In the final analysis, however, productivity within the firm is essentially a management responsibility.

New technology does not of itself raise productivity. Despite the deeply rooted assumption that the introduction of new technology is the job of technical and systems experts, experience has pointed up the crucial role of those who "run" the new technical systems. Those who operate the systems at the grass roots—supervisors and rank-and-file workers—not only want to know how their work life will be affected by the changes; they also want more "say" in the choice of technology and the way it is introduced. Moreover, they have useful ideas to contribute in making change successful.

No part of the productivity equation can be complete without recognition of the strong pressures of international competition: technological advances and the vigorous efforts to penetrate markets once dominated by U.S. industry. Productivity growth among our principal competitor nations has been considerably faster than in the United States. For example, between 1960 and 1976, the average increase in manufacturing productivity per U.S. employee hour was less than 3 percent, as compared with a hefty 8.9 percent for Japan.[6] Although the recent contraction of Japanese export markets, combined with the inability to shed labor, has reduced productivity growth rates there, too, Japan still exceeds us.

Changing Characteristics and Attitudes of the Work Force

The post-World War II baby boom has come of age. Large numbers of young people have moved into the labor force. By the end of the 1980s there will be 60 million workers in the 25-44 age bracket, a 55 percent increase over the 1975 period. The numbers of workers in the 16-24 age group in the 1980s will decline by over one million because of the lower birth rate in the 1960s.[7]

Women constitute about 42 percent of the labor force, and all forecasts predict a steady upward trend in the 1980s. A million and a half women entered the labor force in 1979 alone.[8]

In spite of a temporary relaxation of equal opportunity enforcement in 1982, there will be continuing pressure to employ minorities, who have a disproportionately high rate of unemployment, especially among youth.

In 1959, 29 percent of American employees had an elementary school education or less. By 1979, that group had fallen to 8 percent; it will virtually disappear by the late 1980s. From 1959 to 1977, the percentage of high school graduates in the work force rose from 32 to 42 percent and college graduates, from 10 to 18 percent.[9] The number of college graduates increased from 300,000 to 900,000 per annum during the same period, while the proportion of "professional, technical, and kindred workers" rose 97 percent.[10]

Today, over 21 million Americans have completed four or more years of college, and college enrollment is at an all-time high of 12 million.[11] The number of college students may taper off in the 1980s if federal support is reduced, but it remains a fact that the average length (although not necessarily the average quality) of education among American workers is the highest in the world.

Continuing a major trend of recent decades, a higher proportion of people will move into service industries than into manufacturing and production.

All of these basic changes will profoundly affect the way people view the work environment.

How men and women of the working population approach their roles as productive members of society in the 1980s will be strongly influenced by fundamental changes in attitudes and values. During the 1960s the United States and other Western societies experienced the turmoil of the so-called youth counterculture movement. Feelings among students and others were shaped by the civil rights movement, and later, demands for the rights of all minorities, including women. There was growing concern about the deteriorating environment, convulsive reactions to the Vietnam war, and then Watergate—events which had a substantial impact on social values. The period was marked by both protest and an increasing demand for—as well as tolerance of—self-expression, self-fulfillment, and personal growth.

Even high unemployment and the growing issue of job security in the early 1980s did not destroy these emerging values, which have now taken hold in organizational life, too. The work ethic, a basic belief in the value of work, appears not to have diminished, but many aspects of the relationship between work and the work environment have been brought into question. Younger workers came to feel the incongruence between the freer, more flexible mode of life and expression in the outside community and the authoritative, technocratically controlled workplace. Even long-service career employees felt frustrated that their value to the organization was not recognized.

Ten years ago, 70 percent of industrial workers were willing to accept managerial authority with minor reservations. Today, the reverse finding has emerged: younger, more educated workers resent "authoritarianism." Yet they do *not* oppose the legitimate exercise of authority. This distinction is crucial to worker-manager relations and, therefore, to national productivity in the coming decade.

Since the 1960s, the ruthless pursuit of worker "efficiency," in the industrial engineering sense, has come under increasing attack and "Taylorism" has became a dirty word. A large proportion of the working population viewed work neither as a challenge nor as a personally fulfilling dimension of life, despite the fact that the sheer physical drudgery which characterized so many jobs in earlier times had been engineered out.

General working conditions during the last decade have improved substantially, and so have wages and benefits, even after allowing for inflation and the wrenching effects of the 1981-82 recession. Yet the latest results of the Opinion Research Corporation surveys, which have been conducted over the past 30 years, reveal some striking facts. The extent of job dissatisfaction has increased significantly. Growing discontent is not limited to the factory worker; it is also being expressed by clerical workers, supervisors, and even managers who in previous surveys had shown the opposite trend. With increased automation, the distinctions that once clearly separated salaried and hourly employees are becoming blurred. Both groups value, and expect to get from work, those intrinsic satisfactions (e.g., respect, equity, and responsiveness) that were formerly reserved for managers. All segments of the work force, except those whose very jobs are threatened, are articulating their desire for achievement, recognition, and job challenge.[12]

One examination of more than 300 behavioral science studies dealing with productivity and job satisfaction concluded that increased productivity depends on two propositions:

1. The key to having workers who are both satisfied and productive is *motivation*, that is, arousing and maintaining the will to work effectively. Such workers are productive, not because they are coerced, but because they are committed.

2. Of all of the factors which help to create highly motivated and highly satisfied workers, the principal one appears to be that effective performance be recognized and rewarded—in whatever terms are meaningful to the individual—financial or psychological or both.[13]

Unions continued to assume that their traditional role was to improve the quality of working life of their constituents by bargaining on wages, hours, and general conditions of employment. Although these were central to their institutional function, and especially the issue of job security, some unions were coming to realize that workers were seeking "something more," a "more" which went beyond the normal parameters of collective bargaining.

Given such concerns, just what do workers want today? Essentially they want to feel they are "part of the act." Words like "involvement" and "participation" are overworked, but their substance has real meaning to millions. Over the past ten years there has been a remarkable rise in employee expectations regarding participation in decisions that affect their jobs. In 1977, 54 percent said they had a *right* to take part in decisions affecting their jobs (62 percent of younger workers expressed this view).[14]

The desire is specific to the immediate arena of the job and work-related issues. It does not include executive decisions or corporation-wide decisions extending beyond the individual employee's ken or vision of the organization, except in depressed industries, such as electronics, textiles, automobiles, and steel, where there is a ground swell of resentment against plant closures and the "outsourcing" of production to cheap-labor areas and foreign subsidiaries.

Workers want a say in decisions affecting their *immediate* work environment and the way in which jobs are designed and job changes are introduced. They want to know what the organization's goals are and how managerial authority is exercised in the office and on the shop floor. They want the basic information required to do an effective job. In fact, in a recent national survey, a majority of workers reported that they do not receive enough information to perform their jobs.[15] They want more sensitive and humane supervision, not in the old be-nice-to-people "human relations" sense, but in the sense of helping to develop their potential capabilities and use their talents.

THE CASE FOR INNOVATION IN THE MANAGEMENT OF PEOPLE

The Public Interest

The public has a manifest interest in productivity. The vigor of the economy, the real prices of goods and services, inflation, unemployment, and the flow of tax revenues are linked to productivity growth. For ordinary consumers, what is at stake is the hope of getting more value for their money—a better standard of living.

Unlike the governments of Western Europe or Japan, however, the United States government has not told management how to utilize human resources at the workplace. It has set certain minimum standards—for example, through minimum wage legislation. It has set up a structure for Social Security payments. Federal and state governments have placed restrictions on child labor and maximum hours of work, especially for women. The right of employees to join together, be recognized, and bargain collectively on the issues of wages, hours, and conditions of employment has been established as a matter of national policy by the National Labor Relations Act.

During the 1970s industry was subject to an increasing amount of legislation relating to minorities, women, older workers, veterans, and the disabled. Occupational health and safety, pensions, environmental protection, age discrimination and equal employment enforcement, and the privacy of employee information all emerged as major thrusts affecting people at work. Currently the federal government is attempting to sharply reduce regulations in these areas, relegating some of them to state and local authorities. As a result, the responsibility of meeting the emerging needs of workers will increasingly fall on management and unions. Successful work-innovation programs already "on stream" in many companies strongly suggest healthy directions for managements themselves and for new forms of labor-management cooperation, independent of government regulation.

Employers' Interests—Management Philosophy and Practice to Date

A more educated work force, with changing values toward work, and the rising cost of wages and benefits under conditions of rampant inflation and high interest rates have combined to create the economic necessity for management to secure a better return on both the capital and the human investment. Progressive managers realize they can no longer afford to rely on old methods and old assumptions.

The failure of management to adopt work innovations can only result in a continuation of, perhaps an increase in, the negative behaviors of workers that have plagued many organizations in the past. These negative behaviors include:

☐ Absenteeism, disruptive to work scheduling and output

☐ Turnover, with replacement and training costs

☐ Avoidable grievances, accompanied by stress, anxiety, and interpersonal friction

☐ Defiance of rules, policies, or authority

☐ Militant activity, such as slowdowns, stoppages, or extended strikes, often in defiance of union orders

☐ Sabotage of equipment, materials, or products

☐ Theft or dishonesty in dealings with co-workers or customers

☐ Morale-related accidents and injuries in the workplace

☐ Waste of materials, supplies, equipment, space, or energy

☐ Conflict with co-workers or supervisors

☐ Poor quality of output

☐ Hostility or indifference toward the goals of the organization

Practical work-design programs already carried out in many organizations demonstrate that the negative behaviors cited above can be reversed, with substantial improvements in productivity. The potential benefits for policy makers in the 1980s include:

☐ Better communications and information sharing

☐ Improved quality of service

☐ More effective management and management style

☐ Improved safety and health in the workplace

☐ Better accommodation to technical change

☐ Development of new problem-solving capabilities

☐ Regained employee allegiance and loyalty

☐ Greater employee involvement in the affairs of the employer and the union

☐ Reduction in the number of supervisors required

☐ Diminished counterproductive behaviors and reduction of grievances, work stoppages, strikes, and industrial sabotage

☐ Increased job satisfaction

☐ Human growth and development

☐ Better competitive position both in the United States and abroad

A growing number of U.S. managements no longer allow the working environment to *just happen*. They manage that sector of their business as deliberately as they manage markets, finance, costs, and assets. They have discovered how large a multiplier effect a turned-on, released, and cooperative work force can have on production.

In the past, organizations have focused more on the economics of work than on its psychosocial aspects. This preoccupation tended to detract from quality-of-working-life issues that in the long run may prove to offer more permanent solutions to the problem of rising labor costs.

Top management's commitment to the "human" side of the enterprise has often been solemnly promised in pronouncements which make good press. Executives may even believe what they say; but men and women at the lower levels of large complex organizations believe that the only thing that counts with those on top is making the numbers look good on the balance sheet—and in the short term, at that.

Work innovations have revealed the existence of superfluous layers of supervision and management in the organization. This layering slows decision making, frustrates innovation and change, diminishes the need or desire for risk taking, inflates costs, and deflates results. Extreme functional specialization, together with growing reliance on legal and financial expertise, creates additional obstacles. Viewed from below, there appears to be little hope that the voices of ordinary workers can be heard.

Chief executive officers have not been trained to share power with those in the lower ranks. Many industries are highly centralized and tightly controlled at the top. The steel industry, for example, needed an economic crisis and pressure from the union to move it toward a more participative mode of operation through the realization that management alone cannot turn the ship around.

Once management has recognized the realities of the world of work, the issue then becomes how well it can identify, anticipate, and address these changing values as they surface, or before they surface, in the organization.

New strategies and new policies will bring the modern enterprise into the 1980s with a clearer vision of the importance of employees and a positive response to their newly felt rights and responsibilities. Considerable management training will be needed, along with new concepts and styles of supervision, less autocracy, the skill to manage dissent, and the patience and confidence to expose managerial power to criticism without unreasonable fear of loss of control. The stimulus will come partly from increased factory and office costs, partly from the changes in ideology and intention, and perhaps most of all from the recognition that in order to compete profitably in the market of the 1980s, it will be essential to have a stable, competitive, dedicated, and committed work force, free of the resentments which fester in a poor working environment. The goal for management should be to make positive use of new employee values rather than to temporize and be forced to take reactive and resistive postures.

The Changing Role of Unions

Less than 25 percent of the American work force is represented by unions.[16] Nevertheless, the union as an institution is of enormous importance to the economy. Most of the large industries have contractual agreements which often set the pace for wages, hours, and conditions of employment. Also, there are signs that union representation may increase substantially in the 1980s, especially in the government sector, the not-for-profit sector, among nonprofessional white-collar employees, and in the professions.

Until recently most unions have stuck to traditional bread-and-butter issues in bargaining with management. They continue to assume an adversary role, making demands in behalf of their constituents and reacting to management behaviors which union officials feel are not in the best interests of those whom they represent. In the depressed industries, the unions have been put on the defensive. With many thousands of union members laid off in the recession, the officials have had to reopen contracts and make wage and benefit concessions to save the economic health of an industry. Job security has become the overwhelming issue.

However, unions are coming to realize that their members, while expecting unions to continue their economic and security functions, are also looking for greater involvement in decisions affecting their role at work, including the way jobs are defined and the way work is performed. Traditional contracts do not provide mechanisms for such participation.

Increased worker involvement requires a new kind of union-management cooperation. Unions have shied away from management programs designed to increase worker motivation, viewing them as threats to the "political" integrity of the union—as efforts to steal worker loyalty and even to break the union. Even when management expresses interest in employee-involvement projects, unions have been concerned about management's failure to enunciate and carry through a truly participative philosophy. Many companies, they feel, do not yet accept the union either as an institution or as a partner in promoting the newly emerging worker-involvement process. For that matter, not all unions have accepted the new participative philosophy, either.

But a massive change is underway. Unions and managements are finding problems of mutual interest that need to be solved for the benefit of both. Joint health-and-safety programs have been common for some time, as have apprenticeship training and the administration of benefit plans. More recently there have been joint programs in drug abuse and alcoholism rehabilitation, preretirement counseling, new-employee orientation, and many others.

More encouraging still are the cases in which union and management jointly attack the broader issues related to productivity and quality of working life. Both Ford and General Motors have joined with the UAW, for example, as a matter of top corporate and top union policy, in the interest of greater employee involvement, better-quality performance, and lower absenteeism and turnover. In a number of plants the results have been dramatic. Local managers and union leaders have been amazed at the constructive ideas that have "exploded up" from the shop floor and at the workers' assumption of broader responsibility.

WORK INNOVATION EFFORTS TO DATE—
A BRIEF REVIEW

The number, scope, and forms of successful work innovations that can be applied in the 1980s are legion. There is no one "best way." Each effort must be tailored to the particular history and circumstances of a particular organization, not chosen "off the rack." Each must be structured, pragmatic, and cost-effective over the long term (but not at the workers' expense).

Successful work innovations evolve through a process of individual and group interaction, leading to the solution of problems. The character of the process is:

—cooperative, not authoritarian
—evolutionary and changing, not fixed
—open, not rigid
—inventive, not prescriptive
—interpersonal, not mechanistic

The feature that makes work innovations innovative is that they give workers the opportunity to contribute ideas—and sometimes to make decisions—to improve the management of the enterprise. These opportunities come in many shapes and sizes, over a broad spectrum. At one end is the old-fashioned impersonal suggestion system, in which an employee writes out an idea and drops it into a box in the hope of receiving an individual cash award if the idea is adopted and generates savings for the employer. At the other extreme is the producer cooperative, in which every employee holds a sizable direct stake in ownership, takes a share of responsibility for managing the enterprise, and also does an operative's job.

Most of the intermediate points, which are the subject of this report, are characterized as either consultation or shared decision making. Management consults workers when it seeks their advice *before* making a decision about what work should be done and how it should be done. But some employers go further and give workers the power to decide such questions for themselves, to put their decisions into effect, and to take responsibility for the outcome.

The process by which an organization enhances the consultative and decision-making roles of its members is known as organizational development (OD). "Team building" and "consensus management" exemplify the approach. As a rule, OD programs stop short of the rank and file, especially when there is a union—another indication that ordinary workers are not considered an integral part of the organization. When General Motors and the UAW agreed in 1973 to establish union-management programs analogous to the existing OD programs, the new concept was designated quality of work life (QWL). That term has gained currency as a way of distinguishing between the general process of work innovation and any of its particular manifestations.

Programs of work innovation at the immediate job level also differ in that some focus on the design of *individual* jobs, while others focus on the design of *work groups*.

Individual Job Design

Individual job design involves the well-known practices of job rotation, job sharing, job enlargement, and job enrichment. When implemented successfully, they depart fundamentally from the "scientific management" principles set forth by Taylor, Gantt, Gilbreth, and others. Those principles are based on the assumption that work is most productive when broken into simple constituent motions and made repetitive, with minimum need or room for human judgment. In recent years, methods experts and the industrial engineering fraternity have begun questioning this assumption. A leading industrial engineer recently put it this way:

> For years, we industrial engineers have been fragmenting jobs by pursuing the one motion/one job principle and developing the one best way to do a job. In our headlong drive to save seconds, we forgot about the much more important elements of human productivity—the proper utilization of people's abilities.[17]

Job rotation and job enlargement were among the earliest and simplest methods of departing from scientific management principles. Both add greater variety to the task. *Job enlargement* involves "horizontal loading," that is, increasing the number of different elements a worker performs in sequence in the process of making a product or processing information. The cycle of time to perform a task is increased. In *job rotation*, individual tasks remain the same, but the worker rotates through a number of different nonsequential tasks.

Job enrichment, while also designed to provide more variety, stresses "vertical loading" and, in so doing, becomes a genuine form of work innovation. The employee is given greater freedom to make decisions as to the way his or her work is performed. In successful programs, such as that of Citibank's letter of credit department, many of the decision-making responsibilities traditionally assigned to the first-line supervisor have been incorporated in the worker's job.

Many job-enlargement and job-enrichment programs have failed primarily because they were initiated and implemented solely by management and consultants, with little or no input from the workers themselves, in much the same way that methods experts designed jobs according to "rational" engineering principles. Workers perceived such programs as management gimmicks to make them work harder for the same pay.

Group Job Design

Even more promising are the developments in the improvement of productivity and the quality of working life that utilize the potential creative capacities of *work groups*. Again, the successful examples have taken a variety of forms.

The most publicized work-group innovation is the *quality circle*, or QC, which was originated and adopted widely throughout Japanese industry as the quality control circle. The QC—a voluntary group of workers from a single department, usually led by the foreman or a senior worker—concentrates on identifying and solving job-related quality problems (often with the aid of statistical techniques) and on improving production methods, as part of a company-wide effort. It presents its recommendations to upper-level management, which can reject (with reasons), modify, or adopt them. At the same time, it focuses on the development of the workers themselves. Many of the most successful QC programs, in companies such as General Motors, are operated jointly by local managements and unions.

The QC idea has developed in a variety of other forms, called *employee-involvement teams, participation teams, problem-solving groups*, and the like. Some are structured along the lines of quality circles. That is, the group is made up of a supervisor and his or her immediate subordinates meeting periodically to deal with operating problems. Others include employees representing a larger area of a plant or office. The leader might be a supervisor, a trained "facilitator," or, in some known instances, someone elected by the group from its own membership. Not all groups are homogeneous. In a production organization, for example, they may include representatives of support departments, such as engineering, inspection, maintenance, and others.

A more advanced form is the *semiautonomous work group* or self-managing team. In the production area, for example, operations are so grouped as to allow "team" responsibility not only for sharing and integrating the work tasks, but also for assuming "supervisory" responsibilities, such as scheduling, maintenance, inspection, hiring, training, and even discipline. One dramatic consequence has been a reduction in the amount of direct supervision required, thereby cutting overall costs while giving managers more time for long-range planning. The numbers and types of supervisory duties vary greatly from project to project.

Another important structural mechanism used to implement cost-effective programs, as well as to monitor quality-of-working-life and work-innovation programs, has been *labor-management committees*. These committees also vary widely and are found in both unionized and nonunionized organizations; several hundred are functioning today, and they are spreading rapidly. Some have proven highly effective: they have helped to eliminate waste of energy and materials and to improve the relationship between management and labor. Such committees have been adopted company-wide in multiplant organizations, such as General Motors, AT&T, Bank of America, Cummins Engine, Dana Corporation, and Harman International Industries. Entire industries—steel and retail foods, for example—have established some form of the joint labor-management committee approach. So have entire communities, especially those facing serious economic problems. Jamestown, New York; Lock Haven, Pennsylvania; Youngstown, Ohio; and St. Louis, Missouri, are only a few of several dozen examples.

Technology-Centered Designs

New technical designs in plant layout have enhanced the work-team concept. The Volvo Kalmar plant is one of the better-known examples in which a complex product, or a major component of it, is assembled by work teams. Other designs, known as *island systems* or *buffer zones*, eliminate conveyor pacing by creating free space, with buffer inventories between work stations. Properly introduced, these efforts have clearly led to increased job satisfaction, improved quality of performance, upgrading of skills, increased flexibility in the utilization of employees, and a new dimension in the teaching-learning process.

Changing Forms of Compensation

The workplace of the 1980s will call for new forms of compensation and gain sharing more appropriate to changing job responsibilities. The changing technology mentioned above is one factor. The explosive spread of computer systems can be expected not only in data processing and clerical operations, but also in manufacturing, as computer controls are linked to automated machinery and assembly operations. Traditional payment systems, based on hourly rates, make little sense when workers become "managers" and "supervisors" of expensive automated equipment. Many companies, such as IBM, have already adopted weekly or monthly systems of "wage payments." Some analysts believe that even annual salary plans may replace the present hourly rate system. Such changes can have a substantial effect on how people view the meaning of work and on productivity itself. New pay systems eliminate many elements of class differences between managers and the managed, between white-collar and blue-collar employees. Greater continuity of income and economic security will result. Another factor affecting methods of compensation is that increased responsibility for production will be vested in the work group rather than in the individual worker. Pay systems need to be designed to reinforce group cooperation, sustained effort, and teamwork. Plant-wide and company-wide reward systems can add a valuable incentive for improved teamwork. Innovative executives who have anticipated this trend and its potential benefit for their own business, have introduced new forms of organization-wide economic gain sharing and achieved both greater efficiency and a better quality of working life.

These new forms of compensation are beyond the speculative or theoretical stage. For example, the *Scanlon Plan* has been used successfully for over 40 years. Unlike traditional group incentive plans, which are based primarily on the quantity of products or services produced, the Scanlon Plan is based on cost savings measured by the ratio of payroll costs to sales. The creativity of the group, often an entire plant, is harnessed toward reducing costs. Joint labor-management consultative committees are an integral part of the Scanlon Plan.

The *Rucker Plan*® differs from the Scanlon Plan in that it measures value added rather than sales—that is, material costs are deducted from sales in the formula. Another Scanlon variant, known as *Improshare,*® uses past average productivity as the measurement base. It defines productivity as the total worker hours, indirect as well as direct, required to produce a unit of product or service.

New Work Schedules

Certain new work schedules also increase the range of decisions in which workers play a part, although much less dramatically than the other work innovations described above. Examples of these new work schedules are:

☐ *flexitime*, in which employees can choose their own work hours, subject to restrictions on the length of the day that must be worked, the amount of notice an employee must give on a possible schedule change, and the number of hours of required "core time," when all employees must be in the workplace.

☐ *permanent part-time employment*, in which both job and worker are expected to be part time for a long time; both are regular and stable.

☐ *job sharing*, in which two or more part-time workers share the same full-time job.

☐ *work sharing*, in which a group of workers accept a cut in hours and pay in order to prevent layoffs.

(For a fuller discussion, see Work in America Institute's 1981 policy study report, *New Work Schedules for a Changing Society*.)[18]

Extent of Usage

Firm data on the spread of the various forms of work innovations are hard to come by. However, expert observers place the number of employers with quality circles between 400 and 1,000, bearing in mind that any particular employer may have as many as 300 QCs. The number of employers with self-managing work teams probably does not exceed 100, but again, each employer may have a large number of them. Employers with Scanlon, Rucker, or Improshare plans probably number about 500. Other ad hoc QWL programs exist in large but uncertain numbers. At any rate, the movement toward work innovation has already grown large enough to develop a powerful gravitational field. The table on pages 36-37 lists 200 major employers that have reported on their use of work innovations. It represents only the tip of the iceberg.

NOTES

1. U.S. Congress, Congressional Budget Office, *The Productivity Problem: Alternatives for Action* (Washington, D.C.: U.S. Government Printing Office, 1981), p. 1.

Work Innovators

A
Abbott Laboratories
Accurate Die Casting Company
Allied Tube & Conduit Company
American Airlines, Inc.
American Biltrite Company
American Can Company
American Standard, Inc.
American Telephone & Telegraph
    Company
American Valve & Hydrant Manufac-
    turing Company
American Velvet Company
AMP, Inc.
APA Transport
Atlas Powder Company
Atwood Vacuum Machine Company

B
Babcock & Wilcox Company, Inc.
Baltimore Gas & Electric Company
Bankers Trust Company
Barnes Hospital
R.G. Barry Corporation
Beatrice Foods Company
Bechtel Power Corporation
Bendix Corporation
Bethlehem Steel Corporation
Black Clawson Company
Black & Decker Manufacturing
    Company
Borg & Beck Division, Borg-Warner
    Corporation
Buck Knives, Inc.
Butler Manufacturing Company

C
Caterpillar Tractor Company
Century Brass Works, Inc.
Champion International Corporation
Champion Spark Plug Company
Chicago and Northwest Transporta-
    tion Company
Chicago Title Insurance Company
Citibank, N.A.
Coats & Clark Inc., Crown Fastener
    Division
Columbus Auto Parts Company
Continental Can Company
Continental Illinois Corporation
Control Data Corporation
Adolph Coors Company
Cordis Dow Corporation
Corning Glass Works
Corry Jamestown
Cryovac Division, W.R. Grace
    Company
Cummins Engine Company, Inc.
Cyclops Corporation

D
Dana Corporation
DeSoto, Inc.
Detroit Edison Company
Dexter Corporation
Donnelly Mirrors, Inc.
Dover Elevator Company
E.I. DuPont de Nemours
Durability Interiors Company

E
Eaton Corporation
Emery Air Freight Corporation
The Equitable Life Assurance Society
    of the United States

F
Fairchild Camera & Instrument
    Corporation
Federal Products Corporation
Fenton Art Glass Company
Firestone Tire & Rubber Company
Ford Motor Company

G
General Electric Company
General Foods Corporation
General Motors Corporation
Gould, Inc.
Graphic Controls Corporation
Grove Valve & Regulator Company
Guardian Life Insurance Company
    of America

H
Hackney & Sons, Inc.
Harley-Davidson Motor Company,
    Inc.
Harman International Industries
Harwood Manufacturing Company
H.J. Heinz Company
Hercules Products
Herman Miller, Inc.
Hewlett-Packard Company
Hinderliter Energy Equipment
    Corporation
Honeywell, Inc.
Hughes Aircraft Company
Huron Machine Products

I
ICI Americas Inc.
Industrial Fabricators, Inc.
International Bank for Reconstruc-
    tion & Development (World Bank)
International Business Machines
    Corporation

J
Jamestown, New York, Labor-
    Management Cooperation Project
Johnson & Johnson
Jones & Laughlin Steel Corporation

K Kaiser Aluminum & Chemical
  Corporation
Kaiser Steel Corporation
Koss Corporation

L James B. Lansing Sound, Inc.
La Pointe Machine Tool Company
Levi Strauss & Company
Lincoln Electric Company
Lincoln National Life Insurance
  Company
Litton Industries, Inc.
Lockheed Corporation

M Malden Mills
Manufacturers Hanover Trust Corp.
Martin-Marietta Corporation
McCormick & Company, Inc.
McDonnell Douglas Electronics
  Corporation
Mead Corporation
Mercury Marine
Michigan Bell Telephone Company
Midland-Ross Corporation
Milwaukee Journal
Milwaukee Road
Minneapolis Star & Tribune
Minnesota Mining & Manufacturing
  Company
Monsanto Company
Morse Chain Division, Borg-Warner
  Corporation
Motorola, Inc.
Mount Sinai Medical Center

N Non-Linear Systems, Inc.
Norfolk Naval Shipyard
North American Philips Corporation
Northern Natural Gas Company
Northrop Corporation
Norton Company
Nucor Corporation

O Olin Corporation

P Parker Pen Company
J.C. Penney Company, Inc.
Pentel of America
People Express
Philadelphia Steel & Wire
  Corporation
Phillips Petroleum Company
Polaroid Corporation
PPG Industries, Inc.
Prestolite Company

Preston Trucking Company
Procter & Gamble Company
Prudential Insurance Company
  of America

R RCA Corporation, Picture Tube
  Division
Roth Packing Company
Retail Food Industry Joint Labor-
  Management Committee
RM Friction Materials
Rockwell International Corporation
Rocky Mountain Data Systems

S SAGA Corporation
San Francisco Work Improvement
  Project
Shell Canada, Ltd.
Sherex Chemical Company, Inc.
Sherwin-Williams Corporation
Social Security Administration
Sperry Univac
Sperry Vickers
A.E. Staley Manufacturing Company
Sun Company, Inc.

T Tampa Electric Company
The Tanner Companies
Teledyne, Inc.
Texas Instruments, Inc.
Thor Electronics of California
Travelers Insurance Company
TRW, Inc.

U Universal Machine Company
U.S. Air Force—Ogden Air Logistics
  Center
U.S. Air Force—Oklahoma City Air
  Logistics Center

V Valspar Corporation
Verbatim Corporation

W Walgreen Company
Westinghouse Electric Corporation
Andrew Wilson Company
World Bank

X Xerox Corporation

Z Zilog, Inc.

2. "Productivity Data," *Monthly Labor Review*, May 1982, p. 100.

3. National Center for Productivity and Quality of Working Life, *1977 Annual Report* (Washington, D.C.: U.S. Government Printing Office, 1977), p. 36.

4. Edward F. Denison, *Accounting for Slower Economic Growth: The United States in the 1970s* (Washington, D.C.: The Brookings Institution, 1979), p. 96.

5. Jerome M. Rosow, "Solving the Human Equation in the Productivity Puzzle," *Management Review*, August 1977, p. 41.

6. National Center for Productivity and Quality of Working Life, *1977 Annual Report* (Washington, D.C.: U.S. Government Printing Office, 1977), p. 36.

7. Richard B. Freeman, "The Work Force of the Future: An Overview," in *Work in America: The Decade Ahead*, edited by Clark Kerr and Jerome M. Rosow. Van Nostrand Reinhold/Work in America Institute Series (New York; Van Nostrand Reinhold, 1979), p. 74.

8. U.S. Department of Commerce, Bureau of the Census, *Statistical Abstract of the United States, 1981* (Washington, D.C.: U.S. Government Printing Office, 1981), p. 380.

9. Anne McDougall Young, *Trends in Educational Attainment of Workers, March 1979*, Special Labor Force Report 240 (Washington, D.C.: U.S. Department of Labor, Bureau of Labor Statistics, 1981), p. A-9.

10. Eli Ginzberg, "The Professionalization of the U.S. Labor Force," *Scientific American*, March 1979, p. 49.

11. U.S. Department of Commerce, Bureau of the Census, *Provisional Estimates of Social, Economic, and Housing Characteristics, 1980* (Washington, D.C.: U.S. Government Printing Office, 1982).

U.S. Department of Commerce, Bureau of the Census, *Statistical Abstract of the United States, 1981* (Washington, D.C.: U.S. Government Printing Office, 1981), p. 158.

12. Opinion Research Corporation, *Changing Employee Values in America: Strategies for 1980 and Beyond in Human Resource Planning* (Princeton, N.J.: Opinion Research Corporation, 1980).

13. Raymond A. Katzell, Daniel Yankelovich, and others, *Work, Productivity and Job Satisfaction* (New York: The Psychological Corporation, a subsidiary of Harcourt Brace Jovanovich, 1975), p. 26.

14. Yankelovich, Skelly, and White, "The New Worker," Corporate Priorities, *Briefings for Management*, November 16, 1977.

15. Robert P. Quinn and Graham L. Staines, *The 1977 Quality of Employment Survey* (Ann Arbor, Mich.: Institute for Social Research, The University of Michigan, 1979).

16. U.S. Department of Labor, *Directory of Labor Organizations, 1981* (Washington, D.C.: U.S. Government Printing Office, 1981).

17. Statement made to Robert H. Guest, associate director of the policy study "Productivity through Work Innovations."

18. Work in America Institute, *New Work Schedules for a Changing Society* (New York: Work in America Institute, 1981).

2.
The Employer's Perspective

The decision to innovate is a management decision to be made by the working directors and elected officers who control the enterprise and hold the keys to its future. This power is usually directed toward immediate business problems involving capital, plant, and equipment; product or service; markets; and investments. For top management, managing the human side of the enterprise is clearly secondary.

However, management can and should oversee all facets of the enterprise. Since the human dimension in productivity is the most sensitive and the most subjective, it requires not less, but more attention than other factors. Work innovations, participation in decision making, employee involvement, and increased human-resource management are goals deserving high priority. Only the employer can make the critical choice between continuing the status quo and engaging in a process of change and development to enhance productivity. Not making a change is a decision in itself: it means that management is content with things as they are or is fearful of risking change.

A decision to go ahead deserves critical evaluation. No intelligent executive would launch a new business strategy without sufficient cause or without assessing the potential risks and benefits. The same kind of careful consideration is required of employers who are interested in work innovations.

In evaluating the pros and cons, employers should consider the fact that work innovation is a long-term, far-reaching process, difficult to achieve and to sustain. It threatens old, established practices and deeply embedded value systems and ventilates corners of the organization that have been sealed off for years. Moreover, its success is dependent on a climate of constant encouragement and reinforcement.

The risks are high: not the direct capital costs, which are comparatively modest, but the indirect costs, in terms of morale, supervisory power, decision-making style, feedback, and exposure. The stakes are high, too. Innovations in the workplace open up an entirely new dimension in organizational effectiveness.

This chapter presents the pros and cons. It cites the gains, the costs, the union relationships, the goals, the linkage to productivity, the significance to product and service and, finally, the negative aspects. Hard dollars-and-cents numbers are not featured here, but the costs and benefits are described and the achievements are illustrated. Many of the profound advantages are spelled out and their translation into real savings in the organization is not too difficult to calculate.

What does it take to engage in this process? Are high profits and steady growth disincentives to change? Clearly not; in fact, a number of leading-edge companies became workplace innovators in order to stay out in front. Is a crisis needed to shake some enterprises out of their lethargy? In some cases, apparently, yes. Necessity is the mother of invention here, too. But a crisis may also mean that time is running out and confidence is low.

If work innovations can save factories from closing, or even save an entire company from bankruptcy, do the benefits require further justification? If these social inventions are critical to the survival of sick companies, are they not equally invaluable to healthy ones? Must an employer wait until the illness is almost beyond treatment before looking for medicine?

What we have learned in these studies of almost a decade of American experience is that workplace innovations give the successful organization a sharp competitive edge, a sustaining life force which goes beyond technology and capital, and an inherent capacity for self-renewal.

An organization decides to go with work innovations when it calculates that the likely benefits outweigh the risks. Since the fear of large-scale change itself is usually the strongest deterrent, it often takes an external threat to tip the scales. A crisis was the precipitating cause in the steel industry, in Ford Motor Company, in Herman Miller, in the rubber industry; an impending crisis, in the Japanese automotive industry of the late 1950s; and a distantly foreseen one, in General Motors. However, some highly successful programs have been the product of long-range vision: those in Hewlett-Packard, Lincoln Electric, Donnelly Mirrors, Procter & Gamble, and the Dana Corporation, among others.

The principal gains to be expected from work innovations are: (1) higher productivity and improved quality of product or service, insofar as those can be achieved through greater application of knowledge, skill, and brainpower by the work force; (2) a lasting improvement in the capabilities of the work force; and (3) an enhancement of employee dignity, self-esteem, and job satisfaction. The principal risks are (1) investment of money, time, and effort; and (2) the emotional strain of getting an organization to behave more participatively. It is up to the employer to weigh the pros and cons as objectively as possible before entering into work innovations, and to reevaluate the program periodically.

RECOMMENDATION
1 ◁◁◁◁◁◁◁◁◁◁◁◁◁◁◁

Before entering into work innovations, and from time to time along the way, the employer should try as objectively as possible to assess and reassess the probable gains and risks and the effect on employees.

Presence of a Union

Many employers who might otherwise be willing to try work innovations abstain because a union is present and they see no way of getting the union to join in. This is no idle fear. Most unions are still suspicious of work innovations and of anyone who advocates them. They have been burned too many times by offers of cooperation.

On the other hand, some of the most hopeful programs now going on are those in which tough, militant unions such as the auto workers, the steel workers, or the communications workers are a partner. Toughness, in fact, is the secret of their freedom to cooperate with management. The stronger a union is, and the more its role is accepted by the employer, the better able it is to assure workers that a given work innovation is legitimate and to the workers' advantage.

Employees in many enterprises live with intense frustration, often without understanding why. In the American culture it is normal for people to want to be members of a prosperous, respected organization and to have a visible part in making it succeed. Employers, however, not only frown on such participation, they actively discourage it by punishing employees for increasing productivity.

The typical reward for helping to reduce unit costs is that employees lose earnings or lose their jobs altogether. The urge to hold out a hand in partnership, and the simultaneous inhibition based on fear that the hand will be slapped, underlie much of the bitterness employees feel toward the employer.

A strong union can provide assurance against the possibility that the employer might try to take advantage of those who engage in work innovations. A weaker union's assurance carries less weight. A union that feels that management is out to dislodge or weaken it cannot afford to offer assurance.

Some employers shy away from joint QWL programs because they believe unions are rigid about work rules, seniority rights, and related issues which might come into question. At GM and Ford, the UAW has shown itself ready to experiment, for example, with pay-for-knowledge in semiautonomous work teams, or with gain-sharing schemes, provided that the collective agreement stands firm until the parties have negotiated the change. Local unions, in any case, are much more flexible than the internationals.

The presence of a union, then, need not bar work innovations and may even abet them, but first the employer and the union must reach an understanding that each side respects the other's responsibilities and rights. Chapters 4 and 6 suggest some ways in which this can be done.

An employer considering the advisability of entering into work innovations should not be deterred by the fact that employees are repre- **RECOMMENDATION** *sented by a union. The presence of* ▷▷▷▷▷▷▷▷▷▷▷▷▷ **2** *a union can increase the chances of success, provided that the union (1) is persuaded that the employer does not intend to weaken or dislodge it, and (2) becomes a full partner in the effort.*

Setting up Goals—for Employers and Employees

Since work innovations compel an organization to change its culture—or, in a green-field site (i.e., a new location), to develop a new culture at odds with American tradition—the organization will tolerate them only as long as they generate net gains for all involved: employer, managers, supervisors, employees, and unions. If any of the parties feel disadvantaged, the whole effort is at risk. It is therefore essential to declare in advance what benefits each party expects to derive.

The employer's declared goals for the process should be realistic, candid, pitched to the long term, and consistent with the goals of the other parties. In collective bargaining, goals are distributive; that is, each side aims to get the best possible share of the pot, short of driving the other side into a permanent adversarial posture. In work innovations, goals are additive; that is, each side seeks benefits that increase the contents of the pot.

Distributive goals are usually thought of in connection with unions or rank-and-file employees, but they figure just as strongly in salary and benefit bargaining between managers and the employer. Yet, while few people regard tough bargaining between manager and employer as incompatible with dedication to the goals of the enterprise, it is generally taken for granted that tough bargaining between workers and employer signifies a fundamental antipathy. The difference in perspective reflects the fact that in the United States managers are considered part of the organization, but workers are not. Thus, when managers and union leaders at national or local level first sit down together to talk about what might be done to improve the operation of the organization, both sides are always surprised to find that the areas of agreement are much larger than the areas of disagreement, and that the disagreements are often due to missing or inaccurate information.

Candor in stating goals helps to generate trust. Employees and unions will not be surprised that the employer wishes to improve productivity and profitability; they assume that no sane top manager would overturn the established ways of running the organization without the hope of economic return. As long as the employer's aims are consistent with benefits for the other parties, it is best to be forthright about them.

Employers should not expect that work innovations will produce miracles, even though they sometimes yield near-miraculous results. When top managers are unfamiliar with the operations side of the enterprise, they tend to believe that the chief obstacle to productivity improvement lies in employee attitudes. Occasionally that belief is correct (although employee attitudes are often conditioned by management behavior). Generally, however, it is management's job to lead the way to higher productivity. The function of work innovations is to maximize cooperation, not make up for managerial deficiencies.

A Japanese writer has clarified in the following paragraph the disparate but interdependent roles of management and the work force in raising productivity:

There are limits to what (quality) circles can do. The abrupt quantum leaps in cost reduction that the Japanese have achieved in industries as diverse as steel and consumer electronics do not result from QC circles. Instead they come from major strategic decisions about new technologies and plants and entirely new ways of producing and delivering a product. At Ricoh, for example, it wasn't a circle that figured out how to redesign the business system by changing the technology, manufacturing, and marketing to completely change the game in plain paper copiers. Nor was it the circles that led to the elimination of inventory (the "Kanban" system) at Toyota. QC circles, composed of workers from a single division, can't come up with these bold strokes. Nor can they replace strategy. . . . They are one of a number of productivity techniques which work best when put together.[1]

The employer should ask, "How can work innovations help us make more productive use of our assets over the long term?" This question implies that work innovations apply to *all* employees, not only to rank and file. The jobs of people at all levels and in all areas of the enterprise interact in subtle and continually changing ways to determine how productively all organizational assets are used. The best gain-sharing plans recognize this fact explicitly in the distribution of gains.

As an example of such interaction, consider the planning and scheduling function in any large enterprise (whether it produces goods or services). Each stage of operations must receive its required ingredients at the right time and place, in the right quantity and quality. Otherwise the entire process is interrupted, no matter how conscientious or skilled the operators may be. But planning and scheduling depend on close cooperation among marketing planners, systems planners, operations engineers, line managers, and supervisors, not to mention operators. Let one group fall down on the job and the entire process falls with it.

It is not uncommon for work innovations to bear fruit quickly —sometimes within a matter of weeks or months—but it may take years for any major organizational change to take root. The value of short-term gains is that they create enthusiasm and stimulate the employer, employees, and the union to stay with the effort when the going gets rough, as it inevitably does. With deep-seated interests involved on all sides, only a novice would anticipate clear sailing.

Fortunately, even innovations that do not generate tangible improvements will usually show some desirable changes in employee-management relations or in organizational climate rather quickly. Costs and benefits should, therefore, be judged over the long term, and the employer should draw up strategy and tactics with that expectation. An enterprise that adopts work innovations in hope of quick gain, and then drops it when the gain does not materialize, is likely to find itself worse off than before. Expectations of fast gains with lasting results constitute a shaky foundation on which to base the decision to innovate.

A long time frame is necessary because work innovations involve change in the culture of an organization. Investments in technology and plant and equipment always involve a long-term commitment, with uneven and often delayed payouts. Investment in human-resource management is a parallel commitment, with similar risks and benefits over many years. A whole new set of customs and traditions have to be learned, whether one begins with a green-field site or with an old, established site. Even in a green-field site, where employees are handpicked for the purpose—the Gaines pet food plant in Topeka is a case in point—there is no escaping this need for time. Lincoln Electric Company says it takes two years to "Lincolnize" an employee. When an organization has to unlearn old habits as well as learn new ones, change is even harder.

The more profound the innovation, the longer it takes. Employers who have introduced semiautonomous work teams have discovered that the teams are incapable of carrying out all the duties prescribed for them in less than three or four years. At some General Motors plants, the local union and management required several months just to agree on a set of principles to guide future quality-of-work-life programs. Scanlon Plans may pay good bonuses within the first year of operation, but are not considered permanent until a second referendum, after one or two years of operation, produces a 90 percent majority.

An employer must therefore ask: Are we prepared to stay with the project for the long pull? How will work innovations help us achieve our long-term goals? Are we prepared for a change of culture? Are we prepared to make long-term commitments of managerial interest and involvement, of staff coordination and training time, and of money?

The employer, in addition to setting goals consistent with

those of employees and unions, will find it necessary to bear the others' goals in mind at all times. When the work flow for a complex product or service is being designed, decisions about the layout of machines, equipment, and materials influence and control the shapes of the jobs to be performed by the operators. At Citibank, for example, "job enlargement" that approached individual self-management was a logical outgrowth of a top-to-bottom change from a function-centered organization to a customer-centered one. Another example: semiautonomous work teams are hobbled if supervisors, quality inspectors, personnel staff, and cost accountants perform their jobs in the traditional manner. Nor can top management use the structure of quality circles as a medium for communicating its position during difficult union-management negotiations.

The parties should declare publicly, in advance, what they hope to gain from work innovations, and their declarations should be mutually consistent. The employer's goals, in particular, should be candid, realistic, and long term.

RECOMMENDATION ▷▷▷▷▷▷▷▷▷▷▷▷▷ 3

Apart from specific performance gains, the employer has a right to expect long-term improvement of the organization's human resources. All successful work innovations bring, to one degree or another, the following benefits:

☐ Fuller understanding of the operation, throughout the organization

☐ Higher application of skill levels

☐ A work force with better and more up-to-date information about the opportunities for productivity improvement

☐ More members of the work force who are willing and able to contribute ideas for improvement

☐ More teamwork

☐ Greater willingness and ability of people to manage themselves

☐ Greater openness to change (whether of product, technology, or organization)

☐ An organization capable of functioning with fewer supervisors and fewer levels of management

☐ A dramatic reduction of distractions and dislocations from avoidable conflicts.

Productivity Improvement. The definition of productivity at the enterprise level has become a search for perfection. As far as employers are concerned, an organization has become more productive when it has increased its ability to compete in a free market by providing customers better value for their money. It is top management that decides what particular mix of performance improvements will enable it to do that—for example, product or service design, dealer service, meeting of delivery schedules, quality standards, or price reductions. Work innovations can help in achieving such performance improvements, often to a degree that spells the difference between survival and disaster.

Too often, employers equate productivity improvement with reduction of labor costs, or of labor costs per unit. Yet, the reduction of labor costs per unit may have a negligible (or even deleterious) effect on long-term productivity. Although cost reductions of any kind are always desirable, except when they weaken the enterprise in some unintended way, labor costs in some industries may constitute as little as 5 to 10 percent of total costs, while the prime opportunities for savings are to be found in energy and material usage. It may be more productive to place low priority on labor costs in order to focus the efforts of workers on the high-priority costs of energy and materials. It is also possible that productivity would benefit even more if cost reductions were balanced with the improvement of other aspects of performance, such as meeting delivery promises or upgrading quality.

RECOMMENDATION 4 ◁◁◁◁◁◁◁◁◁◁◁◁◁◁

The employer should not ask the work force to increase "productivity," as such, but rather to improve particular aspects of performance that are within its capacity.

It is well to reiterate that although work innovations *can* help to improve performance in desired ways, they do so only when properly selected, designed, introduced, and managed. The same is true for new technology or any other method of improving productivity. The special attribute of work innovations is that they enhance the value of other means of improving productivity—that is, they permit the intelligence of many knowledgeable and concerned people to be concentrated on the solution of problems through these other means. For example, at the GM plant in Tarrytown, New York, and the Ford plant in Sharonville, Ohio, the advice of shop-floor workers ensured that major changes in production lines were brought up to speed faster and more smoothly than in the past.[2] In other cases, workers have helped to design the layout of new plants, or to shave estimates for a major bid to a customer.

The manner in which the minds of many ordinary workers can be brought to bear on productivity improvement, resulting in enormous gains through a myriad of small advances, is strikingly illustrated by a recent article in the *New York Times*, which reports that:

> In 1981, Toyota received 1.3 million suggestions, an average of 27 from each company employee. . . . Ninety percent of the suggestions were adopted. . . . In the last six months of 1981, these suggestions helped cut costs by more than $45 million. . . . Over all, it was found that Toyota's output per worker was three times the level at Ford. . . . Between 80 and 85 percent of the Toyotas coming off the final assembly line had no defects. . . . At Ford, cars averaged seven to eight defects each. . . . In another corner of the plant, one man fastens parts to a car chassis with specially designed tools. After the work is done, the chassis is lifted and placed at the next work station by a mechanical contraption that looks ungainly and homemade. In fact, a handful of workers designed the automated transportation system and other improvements, with the aid of company production engineers. Five persons used to be at the work station and now there is one. . .[3]

At Toyota, so drastic an elimination of positions does not cause layoffs because employment is guaranteed.

That work innovations do improve performance is attested by the many employers that have stuck with them through thick and thin. No one can compel an employer to practice participation, which is less "natural" to American managers than the traditional style. Programs of "human relations," of the be-nice-to-people variety, are always among the first things to go when hard times set in. Work innovations, on the other hand, have expanded steadily all through the economic turbulence of the past few years, because they have a tangible payoff to employers and employees. General Motors and Ford Motor Company are the most visible proofs of large-scale improvement. Eight other companies are described in *The Innovative Organization: Productivity Programs in Action* (New York: Pergamon Press, 1982), a companion volume to this policy study report. Many more have been reported elsewhere— Procter & Gamble, General Foods, TRW, Cummins Engine, Donnelly Mirrors, Eaton Corporation, Butler Manufacturing, Mars, Inc.,

Guardian Life Insurance, Prudential Insurance, Jones & Laughlin, Dana Corporation, Parker Pen, Herman Miller, Honeywell, Shell Canada. Many others are not as well known, or as well publicized.

Other employers have been equally successful but discuss their results only off the record. Many hesitate to report "bottom line" figures, either for proprietary reasons or because they are skeptical about linking economic outcomes directly to behavioral changes. Such firms, however, report sizable improvements with respect to absenteeism, turnover, grievances, and labor-management relations. Conflicts diminish; teamwork grows. Many say they find it easier to bring new products and processes on line.

Although improvements of these kinds are frequently categorized as "soft," they have important economic consequences. Absenteeism upsets operating schedules and impairs the quality of the product or service (because work either is omitted, or is performed by people who are less skilled or are annoyed at having been shifted about). Turnover means that more people have to be recruited, oriented, and trained—all costly procedures. It causes work units to operate shorthanded; lowers the average level of skill, experience, and loyalty; and wastes the capital investment in training, skill, and loyalty to the company. Grievances consume the valuable time and energy of managers and supervisors, prevent employee ideas from being applied to operating problems, produce constant friction, and impel employees to make the smallest possible contribution to the success of the work unit. "Softness" applies not to the importance of improvements in these areas but to the difficulty of putting a dollar figure on them.

RECOMMENDATION 5 ◁◁◁◁◁◁◁◁◁◁◁◁◁◁ *Employers should take seriously the so-called "soft" benefits of work innovations. Improvements in absenteeism, turnover, grievance load, employee motivation, and teamwork produce hard economic gains. These become normal standards of performance rather than temporary or one-time gains.*

In a recent study, the U.S. General Accounting Office surveyed 38 gain-sharing plans and reported its findings. Of the 38 firms surveyed, 24 provided financial data. "Savings averaged 17.3 percent at the 13 firms with annual sales of less than $100 million. At the other 11 firms, annual sales were $100 million or greater, and savings averaged 16.4 percent. Among the 24 firms providing financial data, those with a productivity sharing plan in effect the longest showed the best performance. Firms that had plans in operation over 5 years averaged almost 29 percent savings in work-force cost for the most recent 5-year period." In addition, 81 percent of the firms reported improved labor-management relations, 47 percent reported fewer grievances, 36 percent reported less absenteeism and turnover, and 47 percent reported other non-monetary benefits.[4]

Another source of cost saving for the employer is a reduction in the number of supervisory positions required per employee in successful work innovations. Fewer supervisors are needed because rank-and-file employees take on more and more duties classically exercised by supervisors; the number of quality-control staff positions needed are also reduced, as described in the following section. The redeployment of surplus supervisors and professionals is discussed in chapter 5.

Some confusion has prevailed regarding the highly publicized General Motors-UAW national agreement on quality of work life. The fact that the agreement studiously avoids the word "productivity" has led many observers to infer that productivity improvement is either irrelevant or, at best, a happy by-product. Nothing could be further from the truth. Productivity *is* an explicit goal of the agreement, but it appears in words better suited to that company's needs: "reduction in employee absenteeism and turnover," and "improvement in the quality of the products manufactured." The word "productivity" is omitted because in the automotive industry it has the connotation of "speedup," which neither party intends. Line speeds are strictly a matter for collective bargaining.

Improvement in Quality of Product and Service. The economic value of controlling absenteeism and turnover is substantial. The importance of quality of product or service is far greater.

Prior to the triumph of Japanese auto imports and the increase of real fuel prices, the quality of American cars had slipped badly, but without greatly affecting sales. Since the mid-1970s, however, quality and fuel efficiency comparisons have steadily increased the percentage of imports. These features are the result of quality control in both engineering design and the manufacturing process. Shop-floor workers can make a significant contribution to quality control in engineering design (if given the opportunity), but they are *indispensable* to quality control in the manufacturing process.

In automotive manufacture and assembly, production speeds and product mix are predetermined. But productivity is not measured simply by the number of units that come off the line. It is measured by the number of *salable* units. Increasing the proportion of salable units produced at the set line speed (other things being equal) thus increases productivity directly. Suppose the line speed is fixed at 60 cars per hour. Productivity is higher if 50 of those cars are salable than if only 40 are. The nonsalable cars may eventually be made salable, but only through incurring additional costs for rework. If the need for rework is discovered after sale, the warranty cost (and the true loss of productivity) is even higher.

Better quality control by workers in the manufacturing process increases overall productivity sharply in two other ways, too. First, it reduces (sometimes almost eliminates) the need for quality inspectors. Second, it reduces the inventory of parts and components needed to maintain line speed. At Toyota, where workers can stop the line to catch a defect, near-perfect quality-control standards enable them to assemble vehicles with a small fraction of the inventory carried by American manufacturers. Lastly, better quality control increases productivity by increasing demand for the product, which in turn improves economies of scale and directly reduces the unit cost of production.

General Motors reports unequivocally that "workplace innovations lead to improved operating performance and higher levels of quality of work life." Statistical studies of 23 assembly plants show that the GM plants with the highest QWL, as measured by a standard GM QWL questionnaire, also have better product quality, higher customer satisfaction ratings, lower absenteeism, fewer grievances, and lower labor costs. More recent analyses have shown that QWL survey findings in one year reliably predict product quality and customer satisfaction ratings the following year.[5]

What is valid for the automotive industry is also valid for all other firms in the field of complex engineering and high technology. As the United States and the rest of the industrialized world try to catch up with the moving Japanese quality target, quality control will assume greater and greater importance in international competition. In time, supporting parts suppliers and service industries also will feel the impact. Reliance on workers for quality control will continue to grow, no matter how much automation and how many robots are introduced.

It would be misleading, however, to leave the impression that quality control can be delegated to the lowest levels of the organization. On the contrary, it demands top-to-bottom commitment. It therefore requires *work innovations within management*. As the president of Fujitsu America, Inc., has written:

We in Japan . . . . have made quality control a responsibility of top management. In the United States . . . the responsibilities of each department are defined in great detail, but no one seems able to take overall responsibility for quality. . . . Fujitsu America had been making a certain subassembly for Amdahl. . . . We discovered that the subassemblies made by Fujitsu America had 10 times as many defects as those made in Japan. . . . Some people seeing this might be tempted to say that American workers are not as good as Japanese workers. This is not true. Like other workers, they need leadership. So I conducted a meeting once every week and had the American managers discuss their work openly. I encouraged everyone to think through problems together, to an eventual solution. Working together, we found these problems: carelessness; poor handling of delicate parts; inaccurate testing equipment; improper placement of parts to be tested; and, most important, failure of managers to act on the foregoing. In due course, the employees discovered that management had no intention of punishing them. The objective was to identify mistakes and correct them. As a result, our Fujitsu America product became equal to—or even better than—the one made in Japan."[6]

Employers should recognize that quality of product or service is increasingly the key to competitiveness, that quality control is increasingly the key to productivity, and that work innovations are indispensable to both.

RECOMMENDATION ▷▷▷▷▷▷▷▷▷▷▷▷▷ 6

Disadvantages of Work Innovations

Every major change in the workplace is attended by costs, including the costs of designing, introducing, refining, monitoring, and routinizing the change. In the case of work innovations, the direct financial investments are secondary. The main costs are the changes of custom and habit that flow from giving up some of management's prerogatives.

Direct monetary costs for work innovations are incurred for external and internal consultancy and for paid hours devoted to training or other meetings. The sums may be substantial. At Tarrytown, for example, over 3,000 workers received 27 hours each

of training in the methods of problem solving and group dynamics—
81,000 worker-hours of training in total. At the nominal rate of $15
per hour for pay and benefits, the training alone came to over
$1,200,000. As an absolute amount, $1.2 million raises eyebrows.
Set against the cost of any technological change that might have
been needed to accomplish a comparable performance improvement, the number looks much smaller. When measured against the
costs of closing down a 3,000-employee assembly plant, an outcome
which the QWL program prevented, it is a pittance. In fact, it was
an investment the company could not afford to pass up.

In general, the capital costs of work innovations pale in comparison to the costs of other means of accomplishing the same ends.
This new path to productivity has made work innovations especially
attractive to basic industries, such as steel, which find that technological changes require enormous investments, with very low rates
of return. Work innovations, on the other hand, utilize in-place
resources, with limited capital costs.

Many employers dislike the idea of investing heavily in training
workers on the grounds that turnover may wash out the benefits.
The objective may be valid in a traditional organization. However,
when the training supports a well-run QWL program, it has quite
the opposite effect: turnover goes down significantly. Thus the
investment pays off in two ways: performance improves as a result
of training, and the better-trained people identify their careers with
the employer and are less likely to quit.

In general, a double standard prevails when employers appraise
proposals for deploying money. Investments in hard goods of any
kind—buildings, land, equipment, machinery—are *ipso facto* preferable to investments in people, without regard to calculations of
payoff. An objective analysis would probably reveal that there is
no less risk, and no greater frequency of success, in hard investments
than in so-called soft investments, such as planning and people.

Of even greater concern to most employers than the financial
costs are the emotional and subjective problems of yielding what
are regarded as management prerogatives. Some of these "rights"
may be about as valuable as a patient's right not to tell the dentist
which tooth hurts, but they are cherished nonetheless. Such is the
right *not* to consult workers or other subordinates about matters
in which they have more information and insights than their bosses
do, or the right to conceal cost information so that workers have
no idea of the real costs of waste or inefficiency.

All work innovations include consultation as a vital ingredient, although some go well beyond that. Supervisors and managers obligate themselves at least to listen to their employees' unsolicited advice on matters touching the workplace and possibly to encourage them to volunteer advice. They may even seek their advice. They do *not* obligate themselves to *accept* advice, but only to give a reasoned reply. Management fully retains the prerogative of making decisions. Yet, so authoritarian is the managerial tradition that this concession to participation in decision making is usually perceived as a threat to entrenched authority.

In fairness, consultation involves more than seeking advice. It is a waste of time to ask for advice unless one is prepared to provide basic information which bears on the problem at hand. If the employer wishes employees to help in cutting costs, they need much more cost data than they (or, for that matter, their supervisors) generally receive. Some self-managing teams do, in fact, see such figures and use them well to control their departments. And gain-sharing plans require the employer to disclose much more information about revenues and expenses than most organizations are willing to do. Unfortunately, many managers are less willing to share information than they are to share money, because knowledge is power.

Next, the consultative process consumes working hours. Employee groups need time and space to identify problems, dig up facts, argue solutions, present recommendations, receive management responses, and sometimes to implement solutions. Other people may have to cover for them during these periods, and some annoyances may ensue. Beyond that, if supervisors or managers are under daily or hourly pressure to produce, they will not easily find the patience for activities that will bear fruit only weeks or months hence. When managers are working too hard to have time to think, they have no time for consultation with subordinates.

Employers face yet another problem. The rapid reduction of grievances in joint QWL programs is a favorable outcome, but it entails some risk for the employer. One reason the grievance load dwindles is that supervisors and union representatives dissipate frictions as soon as they arise; but, in order for that to happen, the employer must grant supervisors more leeway to make decisions about employee grievances than in the past.

As consultation becomes a way of life, supervisors find they have less and less traditional supervisory work to do. This raises problems of job security, reorientation, and retraining, which, if not resolved, poison the atmosphere. Chapter 5 recommends solutions.

RECOMMENDATION 7 ⟨⟨⟨⟨⟨⟨⟨⟨⟨⟨⟨⟨

Employers should regard the financial costs of work innovations as an investment with the potential for a very high rate of return. They should regard the organizational and personal strains of adapting to work innovations as difficult—although not insuperable—and worth the effort.

NOTES

1. Kenichi Ohmae, "Quality Control Circles: They Work and Don't Work," *Wall Street Journal*, March 29, 1982.

2. Robert Guest, "Tarrytown: Quality of Work Life at a General Motors Plant," and "The Sharonville Story: Worker Involvement at a Ford Motor Company Plant," in *The Innovative Organization: Productivity Programs in Action*, edited by Robert Zager and Michael P. Rosow, Pergamon Press/Work in America Institute Series (New York: Pergamon Press, 1982).

3. "The Company That Stopped Detroit," *The New York Times*, August 21, 1982.

4. U.S. General Accounting Office, *Productivity Sharing Programs: Can They Contribute to Productivity Improvement?* (Washington, D.C.: U.S. General Accounting Office, 1981).

5. D.L. Landen and Howard C. Carlson, "Strategies for Diffusing, Evolving, and Institutionalizing Quality of Work Life," in *The Innovative Organization: Productivity Programs in Action*, edited by Robert Zager and Michael P. Rosow, Pergamon Press/Work in America Institute Series (New York: Pergamon Press, 1982).

6. Norihiko Nakayama, ". . .And Some Japanese Remedies," *New York Times*, March 30, 1980.

3.
The Employees' Perspective

It is no longer necessary to argue the case for work innovations on theoretical grounds. Today there is a wealth of solid evidence to show that workers, by and large, are not only interested in work innovations but that, once having tried them successfully, they make strong efforts to sustain them. The best evidence of their willingness to try innovations is the alacrity with which they actually do so when offered the opportunity. Hundreds of quality-of-work-life programs and literally thousands of quality circles have sprung up, in all sorts of organizations; it is not uncommon to find an individual firm with over 100 quality circles. Fad or not, the point is that workers *do* want to give innovations a try, and most do it without expectation of special monetary reward.

Once employees understand what is involved, they almost unanimously endorse work innovations, because they see them as an avenue to job security, higher earnings, dignity, self-esteem, more satisfying jobs, and a more democratic workplace. Their main concern is that the employer's motives may not be genuine.

Employees Vote for Productivity through Work Innovations

For many years proponents of work innovations have urged employers to try them, and employers have cited four reasons for refusing to do so: (1) employees are not interested in making the employer more profitable; (2) employees have no interest in the job itself, but only in its economic rewards; (3) employees have little to contribute toward improving productivity, except to follow orders; (4) consulting with employees about how work should be done is merely a way of being "nice" to them. The fact that Reason No. 4

contradicts Reasons No. 1 and 2, and that Reason No. 3 is contradicted by the existence of thousands of suggestion plans, somehow has escaped notice.

When employees choose innovation at a green-field site, such as the Gaines pet food plant in Topeka, Kansas, or in a company which has been using work innovations for many years, such as Lincoln Electric or Donnelly Mirrors, doubters are prone to say: "Ah, but this is a self-selected group." That argument is no longer supportable. Innovations have been introduced with success in hundreds of establishments previously operated on a strictly traditional basis. Some of the most successful, in fact, are establishments where labor-management relations have been notoriously poor. In these workplaces, a certain amount of self-selection does occur when employees volunteer to take part in the first pilot projects. But once the pilots have achieved results, almost everyone wants to get into the act. For example, at GM's old assembly plant in Tarrytown, which had been run on authoritarian lines from the beginning and which had a militant union local, over 90 percent of the 3,300 workers volunteered to take part in the QWL program.[1]

Most employees are not sure they will enjoy work innovations until they have experienced them. Then they show their approval in the most tangible manner. The pattern is remarkably consistent across the whole range of innovations, from flexitime, as a relatively easy approach, to Scanlon Plans and semiautonomous work teams, as more complex programs. Absenteeism, turnover, disciplinary problems, grievances—all the main indices of job dissatisfaction—go down sharply.

In a recent study of productivity gain-sharing plans, the General Accounting Office found that labor-management relations improved in over 80 percent of the cases. These plans—Scanlon, Rucker, and Improshare—all require a very large majority vote to be adopted, and the Scanlon Plan, in particular, has to be readopted periodically.

The GAO studied 38 firms with productivity gain-sharing plans in practice. The firms were selected to provide a cross-section with regard to type of plan, size of firm, and length of time in place. The great majority claimed to have realized productivity improvements, and accounting records tended to support this.

Employees' support and approval of the plans were manifested, not only by their vote to keep the plans in operation, but by reduced turnover, absenteeism, and grievances; improvement in labor-management relations; job satisfaction; closer identification with the firm; and less resistance to change.

Employees and union representatives cited increased earnings as the most important reason for the improved climate. Other reasons included better labor-management relations, better communications, a greater voice in management, and increased acceptance of employees' suggestions.[2]

These findings indicate that employees—union and nonunion—are glad to advance their own and their employers' economic well-being in concert and to do so through teamwork, diligence, and inventiveness.

The evidence of what employees are actually doing at firms that practice work innovations bears out the findings of numerous surveys. For example, the University of Michigan's quadrennial national survey of quality of employment shows that while earnings and benefits are, as might be expected, of prime concern to most employees, the interest and challenge of the job are also important to a large percentage of them (including many who have probably never heard the words quality of working life). They also show that about 40 percent of the nation's work force have education beyond the high school diploma; a similar percentage say their jobs do not employ their skills fully or offer sufficient opportunity for advancement. In other words, millions of employees feel they could make a greater contribution on the job and would like to do so.[3]

A majority of workers who responded to the 1980 U.S. Chamber of Commerce survey of workers' attitudes toward productivity stated that they believed it possible to "change workers' attitudes and abilities" to bring about the largest improvement in performance in most companies. Workers were generally optimistic and enthusiastic about doing a good job, the survey concluded, despite their belief that they stand to benefit least from improved productivity and performance.[4]

The regular QWL surveys of General Motors installations find, as noted in chapter 2, that high performance and high QWL consistently go together and that high QWL one year is the best predictor of high performance the next. Other employers with effective quality circles or similar consultative groups get comparable results from their surveys.

Employers should adopt the working hypothesis that most of their employees would, in proper circumstances, respond favorably to the prospect of taking part in work innovations.

RECOMMENDATION
▷▷▷▷▷▷▷▷▷▷▷▷▷ 8

Volunteers Only

Although the experience of other employers and employees is instructive, one can never safely assume that it holds true for all employers or employees. Still, it is reasonable to take as a working hypothesis that most of the employees of any given firm would respond favorably to work innovations under the proper circumstances. Not necessarily all of them would be interested, of course, nor would they all show their interest immediately. Some may not understand what is offered to them, and others may want to see proof before they express approval. At the same time, participation cannot be established by fiat. Peer pressure may induce an employee to join a work group in Japan, but not as a rule in the United States. For these reasons no employee should take part in work innovations except as a volunteer.

Employees volunteer in two different ways. They may accept employment with knowledge that the firm is, or is about to be, fully committed to work innovations. Or current employees in a traditionally managed firm may step forward and take part when the firm decides to innovate.

RECOMMENDATION
9

Since cooperation cannot be commanded, and since there is no reliable way of knowing in advance which individual employees are in favor of work innovations, employers should ensure that all participation is genuinely voluntary.

The Evolution of Work Innovations

Employers start work innovations from many different points and proceed along many different lines, but as they gain experience and confidence, they find themselves gradually adjusting the nature of jobs and the organization in ways that tend toward a common set of features. Eric Trist, a pioneer in the field of work innovations, has listed the common features of jobs in an evolving work-innovation program:

Extrinsic Features	Intrinsic Features
Fair and adequate pay	Variety and challenge
Job security	Continuous learning
Benefits	Discretion, autonomy

Trist has also contrasted the features of traditional organizations with those that evolve in the course of a work-innovation program:

Old Paradigm	New Paradigm
The technological imperative	Joint optimization (of technological and social imperatives)
Man as an extension of the machine	Man as complementary to the machine
Man as an expendable spare part	Man as a resource to be developed
Maximum task breakdown, simple narrow skills	Optimum task grouping, multiple broad skills
External controls (supervisors, specialist staffs, procedures)	Internal controls (self-regulating subsystems)
Tall organization chart, autocratic style	Flat organization chart, participative style
Competition, gamesmanship	Collaboration, collegiality
Organization's purposes only	Members' and society's purposes also
Alienation	Commitment
Low risk taking	Innovation[6]

Advantages of Work Innovations for Employees

The pros and cons of work innovations, from the employee's viewpoint, become apparent when their operation is examined in more detail. An analysis of the most advanced participative mechanism, the self-managing work team, for example, reveals the following.

In a typical engineering production plant, a team of seven to fifteen workers are jointly responsible for running a work unit that would normally be headed by a supervisor. Several different machines and other pieces of equipment are utilized to shape and assemble parts into a small component of the plant's final product. Each member of the group, in the past, was responsible for a single task; the supervisor had the job of making sure that tasks were performed in the right order, to acceptable standards, and on time. Now the group *as an entity* makes key decisions: it plans and schedules the unit's work; monitors quality standards; receives cost records and controls expenses; decides which member is to do which tasks; selects, orients, and trains new members; sets up machines; takes care of housekeeping chores; does minor maintenance; and perhaps resolves disciplinary problems. Each member learns the skills of others in order to gain flexibility for the group—and higher earnings for the individual. A member of management (regardless of title) continues to have ultimate responsibility for the results of the unit, but this person exercises authority through support and assistance rather than command. The group meets regularly to make the necessary daily and longer-term decisions and to thrash out interpersonal frictions. Group members receive extensive training in technical and group-management skills, as well as instruction in economics.

From this description one can visualize how much more varied, challenging, meaningful, and complete a job each worker has than in the past. One can also picture the possibility of increased anxiety, excessive clubbiness, petty frictions, peer restraints on creative individuals, and even intergroup rivalries. These risks are anticipated and met by training and accommodation. On the whole, however, the advantages far outweigh the disadvantages.

Other kinds of work teams may be smaller and have less variety and less opportunity for managing work, but the members still enjoy broader and more interesting jobs than the traditional "one person, one task."

Consultative groups, such as quality circles, do not direct the operations of a work unit as a self-managing team does. Each volunteer still has the same daily task as before. The group meets for one hour each week, although individual members may spend additional personal time on problem-solving activities. Variety and challenge are gained because the members tackle problems that go beyond their immediate confines, because they secure technical advice from experts in the firm, and because they learn to sell solutions to managers with whom they would otherwise have only cursory contacts. As important by-products of these activities, employees and supervisors develop increased mutual respect, trust, friendship, and a sense of shared interest in a larger goal. Supervisors begin to ask for advice, and they gradually feel free to delegate some of the day-to-day responsibilities that ordinarily harass employees and unnecessarily burden the supervisor.

There are also many less-structured forms of QWL programs, in which supervisors call their employees together from time to time to solve operating problems as they arise. Here, again, daily tasks have not changed, but the context has.

Groups established as part of Scanlon-type gain-sharing plans display, as a rule, the characteristics of consultative rather than self-managing teams (Donnelly Mirrors is one exception). In addition, the firm-wide basis of rewards gives a special impetus to teamwork within and between work units.

Employees vote in favor of trying and keeping work innovations for both extrinsic and intrinsic reasons: job security, monetary rewards, more interesting work, opportunity to contribute to the organization, opportunity for personal development, more satisfying relations with supervisors and others, and a more democratic workplace. The extrinsic and intrinsic reasons reinforce each other; they are not mutually exclusive.

Job Security. A frequent precipitating cause for the introduction of work innovations has been a threat to the survival or prosperity of a firm. As previously noted, it often takes such a danger to overcome an employer's reluctance to turn to the work force for help. But even without an external threat, work innovations strengthen the organization's competitiveness by tapping a new source of energy, know-how, and commitment. Employees are always aware of how little of their potential is actually put to work, so they view work innovations as an opportunity to increase job security by helping the enterprise grow stronger. In today's economy, as many basic industries and leading companies face crises of survival, the need for job security has intensified and become a latent force in support of union-management approaches to QWL.

Monetary Rewards. Unrealized cost-saving opportunities exist in every organization, and no one is more conscious of them than the rank and file. Employees see gain-sharing plans as a way to capitalize on these opportunities and share in the resulting savings.

Gains may also be shared through the normal collective bargaining process, without the elaborate structure of a gain-sharing plan. Employees are aware that the company's "ability to pay" is a function of its efficiency, profitability, and growth and that, in a sense, their wages and benefits are dependent on the company's achieving these goals. The sharing formula may be explicit, as in a formal gain-sharing plan, or implicit, as in the wage and benefit structure, but it is always a critical element in employee motivation.

Still another avenue to monetary rewards may be the system of pay-for-knowledge, by which a member of a self-managing team receives a pay increment for each additional work unit skill.

More Interesting Work. Every form of work innovation adds to the interest of the job by expanding its content and variety in one way or another. In the case of consultative groups or ad hoc arrangements, employees have a chance to apply their intelligence and knowledge to problems that affect the work unit as a whole, and to view the unit in the context of the entire organization. Members of self-managing teams broaden their skills, do many different tasks, and share in making decisions for the work unit.

Personal Development. The single-task, fixed-work-station job thwarts employee development. Promotion from one such job to a higher-paid job of the same kind does little to broaden the employee's view or ability to bear greater responsibilities. Work innovations enlarge the individual's scope of thought and action and provide opportunities for a wider range of manual and managerial skills. These open up new avenues for advancement and deservedly raise the employee's self-esteem. Employee desires for self-development and growth have kept pace with their rising expectations in accordance with Maslow's hierarchy of needs.

More Satisfying Relationships. In work innovations, the employee experiences a change in relations with supervisors and staff advisors. More decisions are arrived at jointly instead of coming down the line as orders or instructions. The supervisor becomes more of a team leader, posing problems and receiving suggestions for action, and the reasons behind decisions are freely discussed. Supervisors are more likely to be judged on the basis of the performance of the unit than on maintaining discipline or following rigid plans. Employees relieve the supervisor of day-to-day concerns, and the supervisor becomes a supportive advisor. As planning and coordination improve, daily crises diminish. Staff professionals also act less as the instruments of top management rules and more as consultants to the work group.

Democracy in the Workplace. Some union leaders, including Irving Bluestone, former vice-president of the UAW and director of its GM Department, have espoused work innovations as a means of bringing "democratic values" into the workplace. The word "democracy" is subject to so many meanings that this report must try to use it with precision. The standard meaning, on which everyone agrees (except Communist countries), is a political system in which the governed freely elect their leaders. A few companies in the United States do allow self-managing work teams to elect their own supervisors; consultative groups often select their own leaders too, although such leadership carries little authority. Only in worker-owned enterprises do employees elect top managers. Employers and managers need have no fear, therefore, that work innovations imply a basic change of organization control.

There are two ways, however, in which "democracy" seems apt. First and foremost, it suggests that ordinary people are treated with the dignity befitting free citizens. The hallmark of such respect is that those in power value one's advice. High pay without consultation creates a conflict between the employee's status in the outside world and his or her status at work. Genuine consultation at the workplace always lifts the spirits of employees, as they find they no longer have to "check their brains as they enter."

Employees want a work environment more compatible with that of the community. As citizens, they enjoy the right to vote, to question authority, to dissent openly, to receive due process. Work innovations help to bridge the gap.

Second, "democracy" suggests an atmosphere in which the trappings of power are frowned on. Degrees of rank and authority exist, as they must in any society, but they are regarded as transient, or on temporary loan, rather than as belonging to the individual in power. Some employers that adopt work innovations, therefore, also adopt policies that stress equality of treatment, such as putting all employees on salary, and eliminating discrimination with regard to entryways, parking places, dining rooms, clothing, and the like. In this sense, work innovations do bring democratic values to the workplace.

More to the point, during a time of hardship, employees insist that when sacrifice is necessary it should be shared equally. Managers who call for austerity from the rank and file while maintaining their own benefits and perquisites unimpaired must expect a cynical reaction.

In this sense, work innovations do bring democratic values to the workplace.

ocratic values into the workplace without jeopardizing the authority management needs in order to operate successfully. Indeed, work innovations strengthen employees' regard for management.

RECOMMENDATION ▷▷▷▷▷▷▷▷▷▷▷▷ 10

Disadvantages of Work Innovations for Employees

The flip side of the coin, employees' concerns about work innovations, may also be classified as extrinsic and intrinsic.

As noted earlier in this chapter, employees believe that opening the floodgates to ideas for improving operations will increase productivity, strengthen the employer, and thus bring about greater job security and the possibility of higher earnings. But increased productivity cuts both ways. If business volume does not go up enough to offset the reduction of worker-hours per unit of output, the size of the work force may be reduced. If more efficient work methods are developed, employees performing higher-skilled tasks may become superfluous and face demotion or transfer to less-skilled, more poorly paid jobs. These worries must be allayed before employees will cooperate fully in work innovations. Where there is a union, collective bargaining can erect the needed safeguards. Where employees are not unionized, the employer has the burden of ensuring that their interests are not prejudiced through cooperation.

Intrinsic concerns arise from the change in the nature of work itself. While there may be boredom and unused potential in doing a single task under direct orders from a supervisor, with no responsibility to think about better methods of doing it, there is also less stress and more comfort and familiarity. Work innovations prod employees into becoming more venturesome.

The change is substantial, as employees move from narrow certainty to broad uncertainty. A team, for example, has to decide how the work unit's operation can best be carried out, what role individual members will play, and how the roles will shift over time. There is a new responsibility for using one's mind and for being more flexible about task assignments for the good of the unit. There may also be some disorientation as supervisors ask employees how to do the job and employees ask the supervisor and other experts for help in solving broader problems. Fortunately, these anxieties work themselves out pretty quickly.

Employee Attitudes toward Failure of a Work Innovation

An employer that adopts work innovations with the intention of ignoring employee interests is ill advised. Most employees regard it as legitimate for the company to advance its own interests, but not at their expense. If the true objective is to strengthen the employer's hand against the employees, to reduce employees' earnings or job security, or to undercut a union, the program will boomerang.

Arbitrary discontinuation of a program is perceived as a withdrawal of dignity. If employee attitudes were unfavorable before, they will become far more so, and it may take years to undo the harm. Workers do not like to be manipulated.

This is not to say that failures will not be forgiven. When a program appears to have been well meant but poorly executed, employees may well feel annoyed, but they remain open to the possibility of another legitimate try.

Managers Are Employees, Too

Although this report emphasizes rank-and-file employees and the unions that represent them, it applies with equal force to non-union employees, whether blue collar or white collar; supervisory and managerial workers; and clerical, technical, sales, or professional workers.

Work innovations for management employees, through such programs as organizational development (OD), can make a far greater contribution to productivity improvement than can those for rank-and-file workers, because management has by far the larger role to play in that regard. The experience of outstanding American companies, such as Hewlett-Packard, Intel, and Texas Instruments, not to mention the leading Japanese firms, bears out this contention. The true importance of rank-and-file work innovations is that they encourage and eventually require management to follow suit, whereas OD programs, which are more limited in scope, usually stop short of the bottom level.

Every member of management, including the CEO, is someone's subordinate as well as someone's boss. In the capacity of subordinate, the manager has interests directly comparable to those of a blue-collar worker. The gag about "mushroom management"—"keep 'em in the dark and throw horse manure on them"—is funny because it is so close to the bone. Or consider the popularity of the book *Up the Organization* among managers. The difficulty of treating one's subordinates "participatively" when one is subject to an authoritarian boss can hardly be overestimated, yet this is what is required by many OD programs.

The National Advisory Committee for this study unanimously agreed that the CEO's commitment and example are vital to the achievement of productivity through work innovations. The members also agreed that CEOs who treat their own subordinates participatively are uncommon, although a few make an effort to practice what the Graphic Controls Company has called "consensus management." In such an arrangement, senior managers may have the right to take part in selecting their peers or in recommending one of their peers for promotion.

The president of Intel Corporation has described the process of reaching tough decisions without inhibition by rank:

We are sitting around, about eight people, and we don't agree, of course....They are not timid in arguing about it, but somehow, sooner or later, they have to come out of that room with an architecture (i.e., a basic computer circuitry). And I can't help them. The only thing I can do is to make sure that when they beat each other up, nobody gets overwhelmed by one person being more vocal than the other, and to make sure they understand they have to make a decision.[7]

Several companies are now establishing quality circles for managers and supervisors for reasons similar to those suggested in the above quotation. Other companies, such as General Motors, have set up "collateral organizations" (managerial groups outside of the regular structure of the firm) for consultation on specific subjects.[8]

RECOMMENDATION
11 ◁◁◁◁◁◁◁◁◁◁◁◁◁

The CEO should bear in mind that managers and supervisors are also employees who will find it more natural to treat subordinates participatively if they themselves are treated that way. Managers cannot function well under a double standard.

On balance, the pros for employees far outweigh the cons. Most of the cons are in management's hands—lack of commitment, manipulative purpose, and so on. Employees, by and large, are ready and willing to join in. They are a potential energy source, waiting to be tapped.

NOTES

1. Robert Guest, "Tarrytown: Quality of Work Life at a General Motors Plant," in *The Innovative Organization: Productivity Programs in Action*, edited by Robert Zager and Michael P. Rosow, Pergamon Press/Work in America Institute Series (New York: Pergamon Press, 1982).

2. U.S. General Accounting Office, *Productivity Sharing Programs: Can They Contribute to Productivity Improvement?* (Washington, D.C.: U.S. General Accounting Office, 1981), p. 18.

---

3. Robert P. Quinn and Graham L. Staines, *The 1977 Quality of Employment Survey* (Ann Arbor, Mich.: University of Michigan, Institute for Social Research, 1979).

4. U.S. Chamber of Commerce, *Workers' Attitudes toward Productivity: A New Survey* (Washington, D.C.: U.S. Chamber of Commerce, 1980), pp. 11, 23.

5. Eric Trist, *The Evolution of Socio-Technical Systems: A Conceptual Framework and an Action Research Program*, Issues in the Quality of Working Life, No. 2 (Ontario, Canada: Ontario Ministry of Labour, Ontario Quality of Working Life Centre, 1981), p. 30.

6. Ibid., p. 42.

7. "Talking Business with Grove of Intel: A New Approach to Hierarchy," *The New York Times*, December 23, 1980, p. D2.

8. D.L. Landen and Howard C. Carlson, "Strategies for Diffusing, Evolving, and Institutionalizing Quality of Work Life," in *The Innovative Organization: Productivity Programs in Action*, edited by Robert Zager and Michael P. Rosow, Pergamon Press/Work in America Institute Series (New York: Pergamon Press, 1982).

4.
The Union's Perspective

Labor-management cooperation in the workplace has the same ultimate goal as collective bargaining—to bring democratic values into the workplace. But the *process* of cooperation is distinct from negotiation. Both parties accept this separation as basic to maintaining the right of the union to bargain for wages, hours, and working conditions and to meet its traditional contractual responsibility.

Since work innovations are carried out in the plant, at the construction site, or in the office (occasionally in the office of an international union), it is local unions which are chiefly concerned with them. However, local activities are subject to the influence of international policies and attitudes. The head office may encourage, assist, discourage, prohibit, turn a blind eye, or be totally unaware of unreported local programs. This report, therefore, is directed to union leaders at both levels.

Signs of Change

Since 1970 a remarkable transformation has taken place in the American labor movement's position toward productivity and quality of work life. Local union-management QWL projects have spread faster than anyone would have anticipated. At the international level some of the largest industrial and public-sector unions, which earlier considered productivity a dirty word, now freely acknowledge its importance as a reliable source of real wage gains for their members. Their prime concern today is to ensure that productivity gains are not made at the expense of the work force.

The AFL-CIO itself, which tends to be more conservative than its constituent unions, now takes the QWL movement seriously enough to run courses on the subject for union leaders at the George Meany Center in Silver Springs, Maryland.

Only a few international unions have explicitly committed themselves to QWL, but those few—the United Auto Workers, Communications Workers of America, International Brotherhood of Electrical Workers, and United Steelworkers—together represent almost three million workers. Local programs have also been supported by the United Rubber Workers; United Food and Commercial Workers; Bakery, Confectionary, and Tobacco Workers International; International Union of Electrical Workers; International Woodworkers; Oil, Chemical, and Atomic Workers; American Federation of State, County, and Municipal Employees; National Education Association; and some construction unions.

A few international union leaders disparage work innovations, insisting that more pay and benefits will cure whatever is wrong with current working conditions. They point out that they feel no pressure from the membership to participate in QWL. This is hardly surprising, since the great majority have neither had firsthand experience with work innovations nor even observed them in operation. An interest in work innovations is an acquired taste which, once acquired, endures.

Prior to the early 1970s, the only major form of work innovation in which unions collaborated was the Scanlon Plan, and those efforts were local initiatives that were viewed with neutrality by the international unions. Leaders of the internationals regarded Scanlon as essentially a gain-sharing plan, in which the problem-solving committees were a frill.

Although the union movement took part in the series of national commissions on productivity culminating in the establishment of the National Center for Productivity and Quality of Working Life, it did so more as evidence of good will than of belief in the potential of work innovations. What changed the minds of union leaders was the tangible success of numerous joint local QWL experiments during the 1970s, particularly at General Motors. The lesson was reinforced by the crises in several major industries in which union-management cooperation was notoriously absent. In steel, for example, the initiative for joint QWL projects came from the top leadership of the international union and corporate management, rather than from the grass roots.

Renewed Opportunity for Unions

Today the American labor movement stands at the crossroads. The question is whether to continue on the old road or to turn into a new, less clearly charted road to growth. Union membership as a proportion of the labor force has declined steadily over the past few decades. Today, less than one-fourth of the nonagricultural work force belongs to unions, and the power of the labor movement is recognizably diminished.

In addition, unions have suffered from a crisis of leadership in some internationals; failure to anticipate the changing industrial scene; a broad decline in public support for unions as an institution (together with the decline in confidence in other American institutions, but nonetheless worrisome); and loss of political clout in Washington.

In this bleak and troublesome climate, the AFL-CIO and its executive body have engaged in serious self-criticism and an ongoing debate on the labor movement's options for the future. Some solutions are beginning to take shape under the new leadership of Lane Kirkland, such as opening the leadership of the AFL-CIO to women and minorities and broadening the base of the organization through mergers and other political actions. However, the responsibility for increasing membership and strengthening the organization rests with the leadership of the 30-odd major international unions.

Those internationals which have espoused the concept of QWL are pointing the way toward a resurgence of the labor movement in the 1980s. While continuing to bargain in an adversarial style, they have adopted a parallel track of cooperation. The adversarial tradition is a strong one; unions, it is generally agreed, were created to represent protest. Yet not all traditions are worthy of rigid preservation. Times, circumstances, and cultural changes in society call for new responses from management toward employees and their unions, and also from employees and their unions toward supervision and management. This is especially true for areas of mutual concern which lie outside the hard-core controversial issues of economics and working conditions. These new responses are slowly occurring, but the transition raises legitimate anxieties for both unions and management, to be discussed in this chapter. We recognize the concerns of unions, but we affirm our belief, based now on a decade of experience, that advocacy of work-innovation programs will greatly strengthen the position of unions and help to usher in a new era of responsible and effective union contributions to workers and the economy.

Since work innovations can serve the interests of workers and unions, as well as contribute to the survival and prosperity of employers, unions should take a public position in favor of work innovations in principle, without dropping their guard with respect to specific applications.

RECOMMENDATION ▷▷▷▷▷▷▷▷▷▷▷▷ 12

The QWL developments have occurred just when there is a new urgency for union leadership to break with some traditional attitudes. The decline in union membership and the changing composition and values of the work force are pressing problems that will require unions to reassert their effectiveness. The fact that there are more young people in the work force and that they are better educated than their predecessors has raised higher expectations about work life. If unions are to regain strength, they must respond both to their members' needs and to the goals of society.

QWL programs have opened up new avenues of information for unions about internal company conditions and changed the long-term labor-management relationship. When these programs succeed, they present the opportunity of creating an entirely new and deeper relationship between union members and the employer. For example, when Ford Motor Company holds meetings of managers to review local QWL progress, the UAW vice-president for the Ford Department attends and takes an active part in the discussion. In the other direction, the UAW locals of General Motors have been able to use the services of GM's organizational development specialists to help resolve internal union problems.

Active labor-management cooperation is complex. It involves power, diplomacy, resource utilization, and technology. It threatens vested interests on both sides and is usually difficult to cultivate. It requires such long-term commitment that impatience for results can easily abort the effort. Despite these obstacles, clear and definite gains have been achieved, and can be achieved, by any labor union which makes a real commitment to labor-management cooperation for quality of working life.

Unions which are still debating whether to take a public position in favor of work innovations should take note of the potential advantages for union leadership.

RECOMMENDATION ▷▷▷▷▷▷▷▷▷▷▷▷ 13

a. Furtherance of the union goal of democratizing the work-

place. A fundamental goal of all unions has been to bring democracy to the workplace. Work innovations, by creating a relationship of mutual respect and cooperation between supervisors and union members, go a long way toward the goal.

b. *Direct participation in vital decisions about the enterprise.* Union leadership at the local level becomes a participant in day-to-day decisions intimately related to organizational performance. This can only enhance the involvement, the level of effectiveness, and the end results, as well as add to the economic security of the membership.

c. *Strengthened membership due to greater economic security.* Such a cooperative relationship has a positive effect on sales, profits, and market shares; secures the economic base for the membership; and builds in a growth factor for the unions.

d. *Greater know-how about the operation of the enterprise.* Access to "the books" and deeper involvement in the issues of employment, money, materials, and methods raises the knowledge level of local labor leaders.

e. *Greater openness between management and union.* As management confidence grows and the partnership flourishes, the level of secrecy declines and both unions and workers become more fully aware of the real economics of the organization.

f. *Better QWL for union officials, because there is a better balance between problem solving and adversarial activities.* Committeemen and stewards change their role. Instead of acting as full-time grievance processors, overloaded with nagging, needling, repetitious, and petty issues, they become active participants in problem solving. This transformation tends to heal grievances which have festered within the organization for years and produces instead a new and healthier body politic. The number of grievances and local disputes drops dramatically.

g. *Less troublesome negotiation, ratification, and administration of contracts.* Major labor disputes, work stoppages, slowdowns, and wildcat strikes drop sharply; in many of the more successful programs, they virtually disappear. For example, when the UAW/GM national contract of 1973 was signed, inaugurating the national joint QWL program, only two local contracts had been settled. In 1976 the number was eight. In 1979 it was 54, of which a large majority were in plants with joint QWL committees.[1]

The most promising opportunity for strengthening unions today lies in closer ties with employers on matters of shared concern. The political arena will always be critical for unions as a means of both defending past gains and winning new legislative victories for their members and for all American workers. But political achievements do not, in and of themselves, build membership. A strong active partnership between unions and employers must redound to the benefit of both parties. Labor-management cooperation creates a climate for growth and thus provides a better economic base. As the organization grows, so do employment, membership in the union, and the stature and security of unionism itself. Admittedly, this strategy is heavily dependent upon long-term, laborious, plant-by-plant efforts. Nevertheless, in the long run, the goal—a thriving labor movement in a strong economy—will be well worth the effort required to achieve it.

A Wolf in Sheep's Clothing?

Despite the benefits unions can gain from work-innovation programs, many of them have a deep anxiety about QWL and other programs that increase worker satisfaction. The popularity of QWL programs with many nonunion corporations has reinforced the feeling among union leaders that keeping unions out is the real "hidden agenda" of quality-of-working-life programs.

With 75 percent of all American workers outside the union fold, it would be illogical to expect the QWL movement to be restricted to unionized employers. Employers that have a strong urge to be union-free are willing to invest time and money and even to share power in QWL programs in order to achieve this result. Some union leaders have concluded, therefore, that all quality-of-working-life efforts are contaminated.

This concern is not without substance. The QWL process, in one form or another, is being abused by some employers as a means of achieving what the National Association of Manufacturers terms a "union-free environment." It is indeed feasible to develop these programs on a unilateral basis with or without an antiunion motive.

But unions cannot simply wish the problem away. They can respond to the challenge by altering their organizing techniques. In addition to traditional approaches, the union can educate workers to the realization that a QWL program planned, designed, and effectuated unilaterally by management can be altered or withdrawn unilaterally by management. When the union participates and serves as a countervailing force, workers can feel more assured that the QWL process is not within management's sole control and subject to managerial whim, but is, and will continue to be, an integral part of life at work.

Another deterrent for some union leaders is the fact that management consultants play a part in work innovations. Often the assistance of a third party is indispensable in getting started. Although some consultants engage openly or covertly in antiunion programs, many are genuinely interested in advancing joint union-management QWL. The fact remains that, having been hired by the employer, they may, even unconsciously, give greater weight to the employer's objectives than to the employees' or the union's. That problem can be solved: let the union join with management in selecting, instructing, and paying the consultant.

In any event, there is a serious risk to the union in opposing QWL programs when management is for them. If employees like the results of QWL programs and find that they improve job satisfaction, the union will be tagged as denying employees what they may well desire. And if the QWL program is already in effect, the union may be seen by the employees as a threat to the continuation of the program.

RECOMMENDATION 14 ◁◁◁◁◁◁◁◁◁◁◁◁◁ *The most effective way to counter the abuse of QWL programs by antiunion employers is for unions themselves to take the lead in demanding and implementing QWL programs.*

Where QWL programs are not used blatantly to fight unions, there may still be an underlying union concern about QWL programs, fear that loyalty to the union will be weakened. After all, some say, if workers feel that jobs are satisfying, they will soon come to feel they no longer need the union.

The best answer to this concern about "slipping loyalty" is that actual experience shows it to be unfounded. Improving the quality of employees' work life, under circumstances in which the employees know the union is an integral "mover and doer" in making that improvement possible, strengthens rather than weakens their affinity with the union. Just as workers understand that the union is involved in improving their standard of living and their general working conditions, so do they fully understand and appreciate the union's involvement in helping to create a more satisfying work life. Once a QWL process is implemented, employees who are union members display a deeper sense of commitment to the union. Interestingly, their appreciation for management also seems to be enhanced.

The fear that diminished loyalty to unions may result when workers become more satisfied with their jobs warrants closer examination. The union movement grew out of harsh working conditions that created worker protest and such a strong desire for unions that the risks to the individual worker were considered worth taking. As conditions become less harsh, and as employers discover the economic gains to be derived from more secure employment and from tapping the creativity of their employees through increased participation, unions will have to rely less on a wellspring of employee outrage for their existence. They will need to be identified as a positive force in retaining and extending these benefits.

The position of the union leader as an elected official is not to be taken lightly, however. If there is less traditional protest to represent, will union leaders help or hurt their chances of reelection by backing these new programs that increase job satisfaction? The answer, derived from experience, is reassuring. Local union presidents who have supported QWL programs have won reelection easily, and so has every slate of union officers that has campaigned for reelection on this issue.

Unions have always had to contend with the fact that employers have tried to avoid unions by improving pay and working conditions. But unions have never, therefore, been opposed to better pay and better working conditions. Similarly, it would be futile for unions to oppose programs that increase job satisfaction, improve job content, and give employees greater participation in the workplace. Unions will share the credit and become stronger as they become a recognized force in delivering these benefits to their present and potential members.

Walking the Thin Line

The historical relationship between unions and management in the United States has been adversarial. Bargaining over wages and working conditions, with the strike as an ultimate weapon, requires that an adversarial element continue to exist. The embracing of joint QWL programs, with the degree of cooperation and collaboration required to make them work, means that unions and management have to pursue both adversarial and cooperative modes of behavior.

This duality is admittedly difficult. The role change required is equally difficult for both sides—unions and management—since it is unfamiliar and threatening to both parties. In its new role, the union must cooperate wholeheartedly on QWL programs but at the same time maintain the autonomy necessary for tough negotiation. The thin line between cooperation *with* management and subordination *to* management seems almost invisible to many union leaders.

Management has an equally difficult adjustment to make. Most managers have viewed themselves as wholly responsible for operations and in control. They see workers' contribution to operations and decision making as marginal at best. The thin line for them runs between having control and losing it altogether.

In the face of sharp productivity declines, many employers have taken initial steps toward cooperation. Unions, however, remain somewhat skittish and slow to accept the invitation to participate.

Union attitudes toward labor-management cooperation are based on a mixture of considerations:

☐ Managing the enterprise is management's business, most union leaders believe. Because unions have no real control over final decisions, leaders have felt it best to remain aloof.

☐ The adversarial relationship is long-standing, strong, and prevalent on both sides of the table. Union leaders fear that if they establish too close or compatible a relationship with management, they will lose the support of their constituents.

☐ Most union leaders are reluctant to participate in joint efforts because they do not have the expertise of scientists, engineers, or production managers. In a sense, they feel out of their depth and are fearful of being manipulated into a no-win situation.

☐ Despite new and urgent reasons for considering cooperative efforts—persistent unemployment, foreign imports, crises in major industries—many workers and unions still believe that improving productivity necessarily implies speedup and layoffs.

The union movement is founded on a bedrock of realism, and to move toward cooperation requires a realistic understanding of both the difficulties and the promise these new efforts entail.

But realism about the barriers and the history of militancy on both sides should not blur the very important fact that there are aspects of workplace life that are of concern to both parties. Many subjects of common interest have evolved, which are best approached in a climate of joint, cooperative effort. While these are generally well known, it is useful to take inventory of joint efforts in:

—rehabilitation programs for alcoholics and drug addicts
—health and safety programs
—preretirement counseling programs

—programs of orientation for the newly hired
—long-standing apprenticeship programs
—joint efforts to reduce absenteeism

In none of these joint efforts has one side "climbed into bed" with the other. Everyone has gained. No one has lost. The QWL process falls into the same category, but on a different plane.

A union's lack of experience with the QWL process may nourish a sense of anxiety over the unknown. This element is not a new or unusual phenomenon. Every major unexplored undertaking with management started with an "unknown" (the first joint pension program, for instance), but long years of familiarity with the workplace and the exercise of union leadership are excellent tutors. One of the union's roles is to prod management into new, unexplored areas.

The uncertainties surrounding untried work programs yield to actual experience. A survey of union leaders in 1979 revealed:

> In general, those labor union officials who had had con-
> siderable exposure to QWL improvement efforts were
> much more likely to be positively disposed toward union-
> management cooperation in such programs than were
> those without such experience. . . . For example, the
> experienced group viewed joint labor-management
> committees as a natural extension of the union role and
> a practical tool for implementing QWL, while the group
> with little or no experience tended to be skeptical about
> such committees, and to regard them largely as "window
> dressing."[2]

Issues of controversy continue to be the subject of tough bargaining and occasional crisis. But on matters of shared concern, it makes sense for the parties to work together toward solutions that benefit all equally. If the line to be walked in cooperative effort is thin, it is no less real. It can be walked successfully; it *is* being walked successfully in a growing number of joint work-innovation programs.

Union leaders should study and engage in the QWL process while remaining firmly committed to their basic bargaining role. Tough collective bargaining on issues that divide the parties is entirely compatible with day-to-day cooperation on shared interests.

RECOMMENDATION
▷▷▷▷▷▷▷▷▷▷▷ 15

Basic Union Structure and Role Continue

Another concern of union leaders is the degree to which entry into QWL and other cooperative programs may alter the structure of the union or change its role in ways that are threatening to them.

Experience to date does not disclose any case of a union whose structure is altered by reason of the introduction of the QWL process. Neither the international union nor the local unions are affected in basic structure or day-to-day operation. No change in by-laws or provisions establishing union structure is needed in order to implement the QWL process.

In some locations the union may choose to designate one or more coordinators from within its ranks to become "experts" in its behalf. This, however, has no effect upon the structure of the union or its lines of authority. It is similar to the designation of union benefit-plan representatives or health-and-safety representatives to help manage other programs in which the union is involved jointly with management.

Another question asked by unions is whether the QWL process requires changing the constitutional statement of the aims of the union. Generally, the answer is no. The constitutions of most unions are broad enough to contain the purposes of QWL programs. For example, the aim of improving working conditions has always embraced more than merely increasing physical comfort at work. And the standard constitutional aim of extending democracy into the workplace catches the essence of QWL—that is, the involvement of workers in making decisions with regard to a wide range of subjects related to work life, including the means, methods, and processes of work.

In the case of the United Automobile Workers, this enlargement of aims caused the union to rewrite the preface to its international union constitution. Prior to 1980, the following provision was contained in the UAW constitution:

> The worker does not seek to usurp management's functions or ask for a place on the Board of Directors. . . .

The changes adopted in 1980 eliminated this clause. Among the new provisions are the following:

> Managerial decisions have far-reaching impact upon the quality of life enjoyed by the workers, the family, the

community. Management must recognize that it has a basic responsibility to advance the welfare of the workers and the whole society and not the stockholders alone. It is essential, therefore, that the concerns of workers and of society be taken into account when basic managerial decisions are made.

The structure of work established by management is designed to make the workers an adjunct to the tool rather than its master. This, coupled with the authoritarian climate of the workplace, robs the worker of his dignity as an adult human being. This belies the democratic heritage we cherish as citizens in a society rooted in democratic values.

Essential to the UAW's purpose is to afford the opportunity for workers to master their work environment; to achieve not only improvement in their economic status, but of equal importance, to gain from their labors a greater measure of dignity, self-fulfillment, and self-worth.

The workers must have a voice in their own destiny and the right to participate in making decisions that affect their lives before such decisions are made.[3]

This is the formal statement of one union. Other unions may or may not wish to state such enlarged aims in their formal charters or, if they do so, they may want to give different emphasis to the elements of work-innovation programs.

Unions should be open to the QWL process, since it will not alter either their structure or their constitution.

RECOMMENDATION ▷▷▷▷▷▷▷▷▷▷▷▷ 16

Another question important to the union is the impact of a QWL program on the contract. The first answer is that administration of the contract and involvement in QWL programs run on parallel tracks. Collective bargaining proceeds as in the past. Wage matters are handled in the contract, and if company profits and cost savings are enhanced by the operation of QWL programs, the union may bargain for higher wages as a result.

Procedurally also, it is essential to keep collective bargaining separate from cooperative activities. For example, Joseph Scanlon, originator of the Scanlon Plan, insisted that Scanlon Plan agreements should be embodied in separate documents from those of the regular collective bargaining agreement. Cooperation is a day-to-day thing, in which individuals try to combine their ideas and their research in order to resolve problems of common interest. The participants must be open with one another. In negotiations, on the other hand, the competing parties try to make the best possible deal for themselves, although neither side can afford to go so far as to make the other feel cheated. In order to obtain the best possible deal, each side is required to play up the strength of its own claims and the weakness of its opponent's. The use of ambiguity, hyperbole, and even the white lie, not to mention displays of anger and hostility, are permitted. The theatrics of bargaining are inherent in the process. All of these are anathema to joint problem solving.

A further reason for keeping the two relationships separate is that they run on different schedules. Collective bargaining agreements have fixed dates of expiration after which it is assumed that new agreements will be negotiated to take account of economic and social changes that have occurred in the interim. Most cooperative ventures do not operate against deadlines and, when they do, the deadlines are unlikely to coincide with those of the bargaining agreement.

While this clear separation between the labor agreement and a QWL program remains, it is possible that the introduction of QWL programs may lead to a desire to alter particular union-management contractual practices. For example, the parties may feel it desirable to change from a defined, precise wage classification system to one in which workers are paid for their knowledge and skills. This is a matter for collective bargaining and should be handled accordingly.

The officials responsible for collective bargaining on both sides should be direct participants also in fashioning and carrying out the QWL program. This arrangement permits easier discussion and resolution of collective bargaining issues which may arise out of the QWL process and avoids the need for changes in the union structure.

What about the role of individual union officials? Will the committeeman or steward lose stature as a result of the QWL process? Just as the line supervisor or middle manager is usually the most reluctant member of management to accept the QWL process, so too is the committeeman or steward. Supervisors fear erosion of their authority; union representatives fear an erosion of their usefulness in defending the rights of workers. These worries also are allayed by reality.

The union representative soon learns that work innovations facilitate the quick and satisfactory resolution of workplace problems at the source. Fewer formal grievances need be written because problems are settled as they arise. The role of union representatives is not diminished but enlarged in the context of problem solving. They continue to monitor the labor contract and to protect and advance the rights of workers. In addition, however, they participate in the frequent meetings attendant on the QWL process. Some officials take training in problem solving in order to enhance their performance. Their stature among their constituents grows with the increase in job satisfaction.

Unions should maintain a clear line between QWL and collective bargaining, although certain questions arising out of QWL may require settlement through negotiations. Nevertheless, those union officials who are responsible for collective bargaining should also take part in the QWL process.

RECOMMENDATION ▷▷▷▷▷▷▷▷▷▷▷ 17

Thus, the integrity of the labor contract remains intact. Workplace problems are resolved more promptly and satisfactorily. Management-imposed disciplinary layoffs and discharges decline sharply, and controversial collective bargaining issues are resolved more quickly.

In the early days of joint QWL programs, the separation between QWL and collective bargaining was thought to be hard and fast; indeed, it was made a condition precedent to any agreement. As time went on, however, the parties found that the understanding, trust, and problem-solving skills they had developed made it easier to resolve local bargaining issues. Another point at which the two paths intersected was in the UAW negotiations with Ford and General Motors in 1982.

The prolonged recession of 1981-82 forced many industries into wage concession bargaining. Big, tough, high-wage unions, such as the teamsters, rubber workers, and construction workers, gave up some of their past gains in order to preserve jobs against nonunion and foreign competition. They got little or nothing in return.

The UAW negotiations with Ford and GM took place against a similarly dismal backdrop: severe layoffs, heavy corporate losses, grim prospects for future business, and unrelenting foreign competition. In the case of Ford, wage concessions seemed a foregone conclusion, especially in view of the example set by the Chrysler contract. The open question was: How responsive would Ford be to the needs of the union?

Although the improved mutual understanding, open communications, and day-to-day cooperation of the QWL experience could not cancel out the effects of recession, it did shape the tone and outcome of the bargaining. Controversy was muted; the scope for trade-offs was broadened; and discussions of national issues were less plagued by local disputes and frustrations.

According to objective observers, as well as the bargainers themselves, the negotiations took on the atmosphere of a problem-solving session, in which each side not only struggled to get the best terms for itself but also made genuine efforts to limit the damage to the other. The union accommodated the critical short-term cash-flow needs of the company, while the company granted long-term concessions to the job-security needs of the employees. The media have seized on the short-term concessions, which are more dramatic, but there is reason to believe that over the long term the job security arrangements will prove no less important.

Getting Started

Union involvement in work innovations begins with a firm commitment by the local union top leadership. International backing can help, but is not essential. Without local commitment, the QWL process will not overcome the many obstacles that inevitably confront it.

RECOMMENDATION 18 ◁◁◁◁◁◁◁◁◁◁◁◁◁◁ *In order to establish the legitimacy of work innovations, the union should insist at the outset on being an equal partner with management in all matters pertaining to the project.*

In order for the program to be successful, for the union to believe that its goals and those of management are compatible, and for the venture to start off in a climate of trust, coequal status for the union is essential from the beginning. That is, the union must be equal with management in all decisions relating to the QWL process: design, training of personnel, implementation, follow-through, and evaluation.

Otherwise, the program will soon become the object of suspicion and cynicism and be viewed as a "gimmick" rather than as a substantive activity auguring permanent, constructive change in the climate of the workplace.

Once the top leadership of the union (after consultation with, and education of, the membership) is committed to the initiation of a QWL program, it must prepare carefully for the necessary discussions with management and obtain the voluntary participation of workers in the program. In each case, a careful decision must be made as to which level of top union management needs to be committed, that is, local, regional, international, or some combination of these. There is no set pattern to follow, for each QWL program is unique. The only certainty is that the initiating process will absorb time and energy and will demand patience and forbearance on both sides. The following tactical guides are based on experience.

☐ Discuss the matter with selected colleagues in an effort to broaden the base of agreement.

☐ Convene a meeting or series of meetings of all elected local union leaders to discuss the concept and its purposes and goals.

☐ Invite to an initial meeting or meetings an experienced representative of the international union to assist the local unions in initiating the process.

☐ Invite (with the assistance of the international union) an outside consultant in the field (preferably a union-oriented consultant) to provide an objective view, and to develop understanding by means of experiential exercises.

When local union officials believe that management would like to begin a joint program of work innovations, the union should take practical steps to further the aim. For example:

RECOMMENDATION ▷▷▷▷▷▷▷▷▷▷▷▷ 19

a. *Schedule a joint meeting to "break the ice" and establish a climate of trust.* An outside consultant could be involved, perhaps chairing the meeting as a neutral party. The purpose is to establish in the minds of both parties that areas of common interest can be explored cooperatively, to the advantage of both. Pure collective bargaining issues should not be matters for discussion.

b. *Invite union and management representatives experienced in running QWL programs to attend a joint meeting and make presentations.* This would be most useful if the program they ran was in the same corporation or same industry. Such a presentation by peers can help dispel suspicion and distrust.

RECOMMENDATION
20 ◁◁◁◁◁◁◁◁◁◁◁◁◁◁

When the local union wants to inaugurate a QWL program, but the organization shows no interest, the union should take the initiative in bringing it about.

Two ways of enlisting the organization's interest are:

a. Schedule a nonbargaining meeting with appropriate management representatives and ask them to approach the subject with an open mind and engage in a general discussion.

b. If the unit is part of a multiplant corporation, ask the international union to contact the corporation's central office and elicit its help in persuading local management to explore the subject with the local union.

In view of the many impediments to joint union-management work-innovation programs and the limited amount of experience to date, it is not surprising that some programs have failed to get off the ground or, having gotten off, not flown very far. Among the causes of failure are the following:

☐ Management gets too far ahead in planning the program before calling in the union, or asks the union to endorse a program which management has developed unilaterally. This undermines both the program and the potential role of the union.

☐ One or both of the parties set a price for union cooperation before the program has demonstrated that it will produce results. A major program at Volkswagen in Germany failed because the national union made it uneconomic for management to try an experiment in team assembly. There is no harm in negotiating in advance a *formula* for sharing the gains that accrue, as with Scanlon and similar plans, as opposed to putting a monetary value on anticipated changes before they take place.

☐ The chief executive officer announces that the company will deal directly with employees, over the head of the union.

☐ The representatives of management and/or the union who are selected to sit on key joint committees for the program lack the enthusiasm, understanding, or skill to do the job.

☐ The two parties try to move ahead too quickly, not allowing time for essential planning, for getting acquainted, or for devising careful strategy and tactics.

☐ The parties tackle highly controversial problems before acquiring trust and joint experience.

☐ The parties expect results too soon, and when results do

not materialize quickly, they lose interest in the program.

☐ Union leaders, carried away by zeal, plunge into action before the membership is fully prepared to go along.

☐ Third-party advisors to the project are unable to gain or hold the confidence of union and employees.

When management and union share responsibility for a program of work innovation, it is wise for their leaders to sit down together as partners and examine the many possible hazards that face them. By learning from the mishaps as well as from the achievements of other organizations, they increase the possibilities of their own success.

NOTES

1. "The Quality of Work Life Program at General Motors: An Interview with Irving Bluestone," a video cassette tape produced jointly by Work in America Institute and the Institute for Collective Bargaining, 1980.

2. Paul D. Greenberg and Edward M. Glaser, "Viewpoints of Labor Leaders Regarding Quality of Worklife Improvement Programs," *International Review of Applied Psychology* 30 (1981): 157-175.

3. Irving Bluestone, "The Union and the Quality of Work Life Process," an unpublished paper prepared for Work in America Institute, Inc., September 1981.

5.
The Perspective of
Supervisors and Managers

Although the focus of attention in work innovations is upon the better use of workers' abilities, the main impact of change falls on supervisors and middle managers. Workers experience a broadening of the job but supervisors—unless management takes corrective action—experience a contraction. One employer after another reports that the behavior of supervisors can make or break a QWL program, and yet it is rare to find an organization's strategy for change taking this point fully into account.

The Impact on Supervisors

When chief executive officers give the signal to proceed with a work-innovation program, they may not appreciate what is about to happen on the firing line. The simple issuance of an order cannot reverse a tradition which holds that middle managers and supervisors have been selected, trained, and paid to run their operations by telling subordinates what to do and how to do it. Subordinates may offer advice when asked but, for the most part, bosses are expected to know more than the people they supervise and to be the source of ideas for improvement. Then each level presses on the level below to get the work done.

It is easy for the CEO to say, "We are going to be more participative," but managers and supervisors have heard it all before. They nod their heads and go along, waiting to see whether this time the CEO really means it or whether it will blow over in a few months. Evidence that the program is a reality is furnished when the CEO and the rest of higher management come to grips with the inevitable consequences of work innovation:

☐ The supervisor's job has to be redefined.

☐ The supervisor has to be reoriented and retrained for the new job.

☐ The supervisor's employment security must be confirmed.

☐ The supervisor must be given time for consulting with subordinates, and not be pressed for results that compel him or her to revert to the old methods of giving orders.

The employer's strategy for work innovation should take account of the impact on supervisors by providing for (1) redefinition of the supervisor's job, (2) reorientation and retraining to meet the new job definition, (3) employment security, and (4) relief from pressures that might compel the supervisor to fall back on traditional methods.

RECOMMENDATION ▷▷▷▷▷▷▷▷▷▷▷▷ 21

Redefining the Job

Traditionally, employees are promoted to supervision as a reward for having been better at their jobs than anyone else in the work unit. The ability to get along with people may be a plus, as is "leadership," but too often these qualities are subordinated to technical skills when supervisors are selected, in the belief that leadership can be imparted through training after appointment.

Job knowledge is the key, because the supervisor has to train employees, tell them what to do, see that they do it properly and on time, and review the results.

Work innovations stand this tradition on its head. The new axiom is that workers are the experts in how to get the job done, and, moreover, that they want to do it properly and on time, if only the supervisor will let them. If the new axiom is to be followed, the supervisor ought to be actively seeking the advice of subordinates and delegating responsibility to them.

Clearly, supervisors can run a work unit according to one axiom or the other, but they cannot do both at the same time. Hence the need, first, to be convinced that top management really means it, and then, to learn how and to be reasonably willing to cooperate.

Difficulties arise during the transition from the old mode of operation to the new. Even in a green-field site, where supervisors have been assigned and trained beforehand to do a job that is not traditionally defined, workers only gradually increase their proficiency to the degree that will enable supervisors to push duties down the line. In an established site, supervisors will be extremely reluctant to delegate unless they know what their own new duties are going to be.

TRW, Inc., has drawn up a list of duties for supervisors in plants in which relatively autonomous work teams run the work units. The duties which evolve as the work teams become more proficient involve the following:

KNOW-HOW
(1)   Decision making
(2)   Counseling
(3)   Technical advice
(4)   Mechanical ability
(5)   Knowledge of the manufacturing process
(6)   Goal setting
(7)   People motivation
(8)   Transmission of the team concept
(9)   Development of the team
(10) Conflict resolution

PROBLEM SOLVING
(1)   Negotiating
(2)   Acting as referee
(3)   Personal counseling

ACCOUNTABILITY/RESPONSIBILITY
(1)   Production
(2)   Safety
(3)   Cost reduction
(4)   Capital program
(5)   Meeting attendance
(6)   Knowledge of the TRW Paper Work System[1]

These functions go beyond those of the standard foreman's job. Overall, the job is designed on the principle that the same characteristics that define a good job for a rank-and-file employee are equally valid for a supervisor, namely, (1) use of the employee's full capacities, (2) meaningful work, (3) opportunity to advance according to abilities, (4) pay based on knowledge, and (5) opportunity to control the work environment as far as possible.

It is also essential that members of the work team—and the supervisor's boss—understand clearly how the boundary is drawn between the supervisor's duties and the team's. For example, to what extent will the team be held responsible for its mistakes? On what types of decisions is the team required to consult the supervisor in advance?

The work-team solution is more radical than that needed in the case of quality circles or other types of consultative groups, because the amount of delegation to the latter is smaller. Still, consultative groups can (and some do) develop toward self-management. In such cases, General Motors gives supervisors duties as members of interfunctional business teams.

More commonly, there is a gradual shift. Supervisors find it less necessary to "expedite" work or to "police" employees, because the consultative process itself makes employees more attentive to performance. Moreover, as Robert Guest points out in his report on Tarrytown, the groups' problem-solving efforts begin to relieve some of the day-to-day emergencies that make the supervisor's job so frenetic. This encourages the supervisor to devote more time to planning ahead, to smoothing interactions with other work units, to advising the group, and to learning more about the business as a whole.[2]

The employer should redefine the supervisor's job so as to (1) reduce "policing" and "expediting" duties, (2) increase planning, coordinating, and advisory duties, and (3) make it a satisfying whole job. Then the employer should ensure that everyone concerned understands the new definition.

RECOMMENDATION ▷▷▷▷▷▷▷▷▷▷▷▷ 22

Reorienting and Retraining

The skills and insights needed to carry out the supervisor's new duties have not been provided by previous experience. Culling ideas from subordinates, planning ahead, solving interdepartmental problems, and so on, have been downplayed in the past. Supervisors have to be shown as graphically as possible what they have to do differently in the future.

One particularly effective way of orienting supervisors is to involve them in the process of orienting employees toward the work innovation. This serves several purposes. First, it demonstrates that top management does not intend to bypass them in the effort to form an alliance with the rank and file. Second, it lets them see for themselves exactly what new latitude employees are about to be granted. Third, it familiarizes them with the methods of group problem solving. Fourth, it maintains their self-respect by keeping them a step ahead of their subordinates. Fifth, it maintains their authority by making them the people from whom subordinates learn their new role.

Additional training in interpersonal skills and group dynamics will be helpful to all supervisors. Beyond that, however, the contents and timing of training should be keyed to the particulars of the new job definition, which will vary considerably from one work unit to another, and from one site to another. Frequent interchange of experience with peers and bosses helps to stimulate learning.

Employment Security

The issue of job security for supervisors in connection with work innovations is too often ignored. Since supervisors normally come out of the ranks and are considered superworkers rather than managers, they are only a little less expendable than workers. Anything that smacks of allowing workers to manage themselves is a threat to their livelihood.

On the other hand, giving employees more of a say should enable the firm to operate with fewer supervisors (and quality-control staff). Indeed, some employers enter into work innovations with the stated goal of reducing the supervisory staff by one or two levels. This may be a legitimate objective, but it is hardly calculated to win the enthusiasm of supervisors.

Even if supervisors' jobs are redefined, there may be fewer slots to fill. In addition, some supervisors may not be adaptable enough for retraining.

Employers will want to anticipate these problems and work out practical solutions. Job security, in the sense of a promise of continuation in present responsibilities, is not practical. Employment security—that is, assurance of suitable position and income within the firm—would give evidence of the employer's appreciation of those who help to increase productivity. If employment security is beyond the employer's capability, the employer should at least provide paid outplacement services until the redundant supervisor finds a suitable job elsewhere.

*Since the number of super-
visory positions may be reduced, and
some supervisors may not be capable
of broader responsibilities, the employer should assure all current
supervisors that they will either receive alternative positions of suitable*

RECOMMENDATION
▷▷▷▷▷▷▷▷▷▷▷▷ 23

*status and income within the firm, or be assisted until they find
suitable employment elsewhere.*

Pressure for Results

The acid test of whether the CEO really means to change the organization's culture comes when the supervisor puts the work innovation into effect. How much annoyance does the boss show

☐ when a consultative or self-managing group spends time in meetings?

☐ when the supervisor spends time giving the group information or advice they have requested?

☐ when a group's recommendation requires time and money to follow through?

☐ when the work unit's deliberations cause temporary snags in production?

☐ when a group makes a mistaken decision?

☐ when the supervisor and the group try to keep the innovation going despite a sudden influx of orders needing immediate attention?

Stating these questions makes it clear that unless the supervisor's boss, and the boss's boss, and all of management up the ladder through the CEO, genuinely support work innovation, there is little hope for success at ground level.

Advantages of Work Innovations for Supervisors

It would be erroneous, however, to leave the impression that work innovations offer the supervisor nothing but headaches. They can, in fact, improve the quality of the supervisor's work life considerably. Among the benefits are:

☐ A more stable work force, because absenteeism and turnover are reduced.

☐ Fewer grievances and conflicts, because the new relationship between supervisor and employees allows problems to be resolved quickly and without rancor.

- ☐ More teamwork, because supervisors' and employees' attention is directed toward shared goals.
- ☐ Better employee understanding of what the work unit is trying to accomplish.
- ☐ Greater initiative and more constructive ideas put forward by employees to improve quality and cost-efficiency.
- ☐ Reduction of pressure, as employee initiatives and solutions eliminate obstacles to production.
- ☐ More opportunity to plan ahead, which is made possible by the reduction of pressure and which helps to reduce pressure further.
- ☐ Broader responsibilities and fuller participation in management decision making.

The realization of these benefits does more, in the long run, to win the backing of supervisors than does any amount of indoctrination or exhortation.

The Impact on Managers

When the CEO gives the order to proceed with work innovation, that instruction is useless unless it commits the CEO's personal time and energy as well as the resources of the firm. The CEO must demonstrate to each level of management, in turn, an understanding of the problems work innovation will create for managers, as well as the benefits. The CEO must also indicate that successful management of work innovations is one element of performance to be considered in deciding promotions and increases.

The main problem for a manager, as noted above, is to give up the prerogative of *not* consulting employees about their work. Positively stated, managers must devote time and effort to nurturing a QWL program. They have to allow supervisors greater leeway in dealing with employees. They have to take part in reviewing and replying to recommendations from the work groups and in granting recognition when due.

The CEO and other managers will be called on to demonstrate commitment not only at the inception of the process but for the duration. Work innovations do not have a beginning, middle, and end. Either they become a permanent feature of organization life or they die.

RECOMMENDATION
24 ◁◁◁◁◁◁◁◁◁◁◁◁◁

The CEO should commit personal time and energy not only to launching the work innovation but to ensuring that it becomes a permanent way of life for the organization.

Rewarding Managers

There are two major difficulties in rewarding managers for their performance with respect to work innovations: first, determining how well they have performed; second, favoring behavior oriented to the long term when, in most other respects, the system rewards short-term results.

How should the success of supervisors and managers with work innovations be measured? The answer must be sought in the original definition of goals: not only the employer's goals, but the employees' as well. For example, if the economic goals were to improve the quality of the product or service and to reduce absenteeism and turnover, and if the employees' goal was to develop a more satisfactory working relationship with supervisors, then the appropriate measure is how much progress has been made in those directions. It may not always be clear how to determine checkpoints for long-term improvements, but the effort must be made.

The conflict between the current system of measuring results in a short time frame and the need to take a long-term view of work innovations puts an unfair strain on managers. While the stock market wants to see a flashy bottom line in each quarterly report, the long-run health of the enterprise may be better served by taking low or zero profits for a year or two. The CEO's commitment to work innovation may be measured by willingness to risk the impatience of the shareholders in order to pay greater attention to the development of subordinates.

A number of organizations are experimenting with ways of rewarding managers for long-term performance. One possibility is to base bonuses on results achieved over a period of two years or more, rather than one year. Another is to make stock options exercisable only after a period of, say, five years from the date of the award.

Supervisors and managers are the lifeline of effective organizational performance. However, the fundamental nature of QWL and participation places new stress on managerial and supervisory relations. Many middle managers, in fact, already complain about a sense of powerlessness. One cannot assume, therefore, that either group will get into the swing of things under this new management style.

In order to retain the interest of managers and supervisors, programs should be hypersensitive to their needs; investment and reinvestment in managerial and supervisory training are essential; and readings of supervisory attitudes are required at appropriate intervals. Top management and program leaders should be highly responsive to these attitudes and make appropriate modifications whenever necessary.

NOTES

1. James E. Hamerstone, "Work Teams," an unpublished paper prepared for Work in America Institute, May 1981.

2. Robert Guest, "Tarrytown: Quality of Work Life at a General Motors Plant," in *The Innovative Organization: Productivity Programs in Action*, edited by Robert Zager and Michael P. Rosow, Pergamon Press/Work in America Institute Series (New York: Pergamon Press, 1982).

6.
Preparing, Launching, and Pilot Testing Work Innovations

The process of carrying out a quality-of-work-life effort is arduous and complicated. It requires patience, intelligence, and persistence. It requires a fundamental examination of traditional leadership styles and of the history, culture, and background philosophy of management. Where there is a union, the union's position must be known and understood.

This chapter draws upon many successful and unsuccessful experiences to indicate how change can be introduced at the highest levels of the organization, including top-level policy formulation and statements of a new philosophy, and how joint union-management agreements set the stage for a quality-of-work-life effort. How is a plan developed? How is the intent of a statement of philosophy communicated throughout the organization?

The chapter also explores step by step the process of innovation at the local level, in industrial plants, business offices, and other work units. Paralleling what goes on at headquarters, there needs to be top management and union commitment at the local level and, again, effective communication of the intent to all levels. Should a local effort be mandatory or carried out on a voluntary basis? Should there be a master plan before beginning any kind of effort, or can it be incremental? How much flexibility should there be in setting up the rules of the game and in structuring the project? How is the line drawn between the quality-of-work-life process and collective bargaining agreements?

Structural arrangements are also examined, beginning with the top coordinating group in a given plant or office, the selection and training of coordinators and facilitators, the media for com-

munication, and types of work groups. What role does line supervision play in sustaining any project? How can staff and support
groups be utilized effectively?

The following chapters consider how a work innovation, once
launched, can be diffused into other segments of the organization,
and, of critical importance, how activities can be sustained over
time.

Assessing the Organization and Preparing for Change

Experience has provided some useful warnings about assessing
the desirability of work innovations for a given organization. An
effort at the grass-roots level will suffer in organizations which experience peak-and-dip economic conditions and thus wide fluctuations in the size of the labor force. Also it has been found that high
turnover of management personnel interferes with the sustained
commitment over time which a quality-of-work-life effort requires
from the leadership. When committed leaders leave, there is a gap
in knowledge, experience, and commitment. Often the effort has
to start again from square one. Finally, where labor-management
relations have traditionally been marked by intense mistrust, success
is unlikely—unless a great deal of preliminary rapprochement takes
place. There are rare examples of success in which one or more of
these conditions have existed, but the implementation process took
years to accomplish.

The first step in any kind of work innovation is the approval
and support of the top management of the organization (and of
the union, if one exists). The process affects the entire organization
and requires a change in the way top managers, middle managers,
supervisors, support specialists, and employees deal with one
another. In essence, it involves a downward shift of power relations
among these groups. If management does not comprehend this, it
cannot deal with the stresses and strains that inevitably arise.

Such commitment from the top is not easy. It implies a basic
change from autocratic to participative style. The CEO and CEO's
lieutenants have to be willing to reexamine their management
philosophy and the assumptions they make about themselves and
about the role of people at all levels of the organization.

Commitment to the QWL philosophy begins at the top, with
the CEO's understanding and endorsement. An increasing number
of top executives have learned, either from the press, business
journals, or their human resources staff, something about quality
circles and other developments. To gain further knowledge and

understanding, it is useful to get advice from a respected consultant. Informal face-to-face sessions with other CEOs, from companies that use work innovations, are particularly effective in overcoming skepticism.

Through discussion with top management, the CEO clarifies the kind of change that makes the best sense for the organization. Questions about personal and company values should be explored—questions that go well beyond starting some "programmatic" effort aimed at bringing quick improvement of the company's return on investment.

One CEO, whose company practices QWL throughout, followed the initial discussions with top associates by working up a simple, broad, written statement to the entire management body. The statement outlined the proposed new managerial philosophy, the democratic value system on which it was based, and the fact that the CEO intended it to be followed at all levels of the organization. It included a warning against precipitate action.

Putting a new managerial philosophy into practice requires patience and skill. A statement embracing participation is only a tiny step foward, especially where the tradition of autocratic rule has operated for a long time.

Before an organization sets out on a course of work innovation, the CEO should announce that he or she is committed to that change. This commitment should be for the long term and should include the needed resources.

RECOMMENDATION ▷▷▷▷▷▷▷▷▷▷▷▷ 25

Even where there is commitment, a quality-of-work-life effort is in danger of being perceived to be much like other past human relations programs, that is, a "program" to be installed and run by a staff group or a friendly gesture toward clerical and blue-collar workers. If anything at all has been learned, it is that commitment must be pervasive and that ultimately the effort must be carried out through the entire organization.

If management has already had a favorable experience with organizational development (OD) programs, achieving commitment is easier. The fact that General Motors had had such experience, while Ford and Chrysler had not, was part of the reason why QWL got under way so much earlier at GM. (The other part was UAW Vice-President Irving Bluestone.)

Top management has to be willing to "go public" (within the

company) with a clear-cut statement of readiness to change a traditional management philosophy and value system. This needs to be communicated to all levels. Officers at the upper levels of large organizations do not commit themselves when the chief executive officer merely issues an order for cooperation. Statement of intent must be followed by behavior that manifests a participative approach.

Behavior includes symbolic gestures, especially in the early stages of change. Every act of management is regarded throughout the organization as evidence of policy—even if management has no conscious policy. Thus, the public attendance of top corporate and/or union leaders at the inception of a local QWL program flashes a stronger signal of support than, say, a memo, a telegram, or a phone call. Operational methods and practices also may unwittingly contradict management's words, as, for example, when employees are told that they are trusted to be responsible and to respond to unexpected conditions, and then are asked to obey an operating rule that requires them to phone in for instructions every 15 minutes. Similar confusion arises when a group at a remote site is permitted to use its initiative to devise a variation of a standard practice, and later is criticized by headquarters for having done so.

RECOMMENDATION 26 ◁◁◁◁◁◁◁◁◁◁◁◁◁

Once having declared his or her commitment, the CEO should take the responsibility for ensuring that all levels of management support the proposed change.

The substance of a formal pronouncement at the top should be carefully worked out, lest its intended meaning become distorted as it passes down through the organization. Where such a statement is perceived simply as a ploy to get more work out of people for the same pay, it is bound to fail. The central theme of the CEO's announcement should be the intention to tap the creative abilities of the entire organization for the benefit of the entire organization and to provide everyone with an opportunity to enhance self-esteem and job satisfaction by getting involved in the decision-making process.

As a next step, the organization's "readiness" for change needs to be assessed through a mechanism that can provide an accurate diagnosis of the organization and, at the same time, alert people at all levels to the fact that management has committed itself. For example, in a large steel corporation which participated in a national agreement with the United Steelworkers to institute "participation teams" throughout the industry, top management hired consultants to facilitate the process at the grass-roots levels. But the CEO also wanted to assess communications and work relationships for the entire management staff. Another professional consultant was brought in to conduct a survey carefully tailored to the history and circumstances of the company. More than 90 top corporate and division managers and executives were interviewed, and then a detailed written questionnaire was given to more than 950 managers above the rank of general foreman or section leader.

The results were fed back, not to top corporate management alone, but to all parts of the general office, divisions, and operating units. The survey data were used for many purposes: to identify whether the communications process was functioning smoothly enough to stimulate participation, to identify roadblocks, and to provide managerial personnel in all units with concrete information to assist them in addressing communications and work-relationship problems.

In other words, the survey results served as a springboard for action. And the action to be taken was arrived at, not by dictation from the top, but by involvement of those who had to initiate and carry out changes. Simultaneously a "bottom-up" process was inaugurated through the establishment of "participation teams" among workers, union representatives, and front-line supervisors.

Sources of Help. Having made the commitment to engage in a long-term QWL effort, those responsible for launching it may feel the need for professional help. Such help can come from a number of sources—internal staff professionals, outside consultants, academic researchers, and others.

Large corporations usually have staff persons who are knowledgeable about new developments in the area. Many organizations have been engaged in organizational development programs. Their training facilitators have conducted "team-building" programs as a means of improved group and intraorganizational communications and greater openness and trust. Through experiential exercises they have trained people in group leadership and problem solving. Such training skills are useful and appropriate for introducing QWL changes. (Unfortunately, most OD programs have been limited to management or white-collar personnel. When attempted in unionized plants they have appeared as unilateral management efforts and hence have been met with suspicion from union representatives. The essence of quality circles and other forms of QWL commitment, on the other hand, is that all parties are involved in directing the effort.)

Outside management consultants can be particularly useful in preparing an organization for change. Those who have had extensive experience and understand the principles and philosophy of QWL can, aside from all else, warn an organization about the many mistakes that can occur during the launching period. Using a variety of survey techniques, they can provide diagnostic data which will become important in deciding how the effort should be structured, what people and what functions should be represented in the structure, and where to get started. Outside consultants can help set up workshops and seminars for those requiring orientation and guidance.

It is wise to check thoroughly with previous clients before employing a consultant. The popularity of QWL today has attracted many practitioners who lack in-depth, long-range experience, so that considerable care needs to be taken in making a choice. Some make flashy presentations and promise results that sound impressive but lack substance. They sell a "package" which, by definition, cannot be flexible enough to fit the unique history, background, technology, and personality of an organization.

Other consultants move in to "manage" a program, thus violating the very principles of involvement, participation, and self-direction. By making themselves indispensable, they damage their clients in the long run. Some accept assignments with the express purpose of destroying a union or keeping a union out.

Some large universities have faculty members who are well qualified in the QWL area. They hold conferences periodically for persons from industry, government, and academia to stimulate and educate the participants in the advantages of new forms of work innovation. Many universities have established centers or institutes to keep abreast of developments in the field, to perform research, and to stay current. Members of these centers or institutes are available for consultation.

Choosing a Program. Having decided on work innovation *in principle*, the parties must choose a particular program. The three most prominent categories to draw from are gain-sharing plans, self-managing teams, and consultative groups. The three are not mutually exclusive; the choice depends on circumstances and personal taste. Since consultative groups are the most flexible, the fastest growing, and by far the most popular medium of change, most of this chapter is oriented toward them.

Scanlon, Rucker, and Improshare plans are often a means of reunifying an organization which has been torn apart by an individual or small-group incentive scheme. Since they measure productivity and pay bonuses on the basis of the *entire* organization's performance, blue collar and white collar, they are applied throughout the organization from the beginning. Therefore, they must be totally designed beforehand; but if the process of design is carried out by the entire organization (usually through representatives), it serves to forge a lasting spirit of teamwork. Participants learn more about the business than they could in any other way. Trust is enhanced. Workers are motivated to share responsibility for running the business through consultative groups called departmental suggestion committees.

On the other hand, top management in a gain-sharing plan must be prepared to open the books to all the participants. Without candid, honest reporting of the accounts, the plan cannot work. It also makes participants intensely aware of management mistakes. In addition, the success of the plan hinges on paying fairly steady monthly bonuses. If the enterprise is visibly heading into troubled economic waters, a gain-sharing plan may be hazardous.

Self-managing teams (SMTs), in the right circumstances, achieve very high productivity and very high motivation; but they appear to work best when introduced organization-wide as part of a total organization redesign. So, as in gain-sharing plans, the entire change must be designed in advance, with each team focused on a shared output. Management must commit itself fully to delegating decision-making powers to workers before it has had any experience with such a course. Even if a less-than-complete delegation is made at the beginning, so that the teams may enlarge their responsibilities gradually, management's initial commitment is binding. Despite the outstanding results produced by SMTs in General Motors, General Foods, Procter & Gamble, and elsewhere, not every employer will want to plunge in so far at once. Also SMTs precipitate job-security problems for supervisors almost immediately.

Consultative groups (quality circles, employee involvement groups, labor-management committees, and others) steer a middle path. Although less effective than SMTs or the gain-sharing committee mechanism, they have the virtue of allowing time and space for experiment. They can evolve toward SMTs or remain strictly consultative but undertake increasingly difficult and far-reaching problems. Management has to commit itself to change, but not to a preconceived pace and form of change. Supervisors also have more time to adjust than in an immediate organization-wide change.

Quality circles are especially adaptable, which accounts for much of their rapid spread. They work for blue-collar employees, for clerks, for technicians, for salespeople. They lend themselves to useful hybrids, such as interdepartmental circles, or circles made up of the employees of a company and the employees of one of its suppliers (or one of its principal customers). They facilitate the transfer of expertise from professional staff to workers and supervisors.

Many warnings have been sounded about the faddish and mechanistic introduction of quality circles. The warnings are in order. However, if an employer adopts quality circles with a full understanding that they require large-scale cultural change in the organization, the advantages of this technique are very great.

RECOMMENDATION
27 ◁◁◁◁◁◁◁◁◁◁◁◁◁◁

Before undertaking action at the grass roots, the CEO should assess the degree of readiness for change in the various parts of the organization and should, if feasible, use the findings as a basis for initial change activities. The employer should also be careful to choose a form of work innovation that is compatible with the organization's true nature, because a mechanism that requires sharing more information or more decision-making power than management can live with is doomed to failure.

Implementing Change—Starting at the Top

The more successful among the large corporate organizations issue a written policy statement of their intent to support QWL (or its equivalent name, as they choose). The statement emphasizes the importance of participation of *all* employees, managerial and nonmanagerial alike. It stresses creativity and improvement in the entire work climate. It declares the objective of achieving the multiple goals of the organization *and* of its people. As a policy statement, it does not spell out procedural details.

Because flexibility in implementation is a key ingredient of any effort, the procedural details should be left up to subordinate units—individual subsidiary companies, divisions, or departments.

Unions, where they exist, are essential partners in implementing work innovations. Much of what was said about management applies equally to union leadership: top-level commitment, explicitly stated and followed through to the local level. As indicated earlier, the union movement as a whole has yet to embrace the quality-of-work-life concept, but the idea and its implementation are growing, especially in the strong national unions, such as automobiles, steel, telecommunications, and parts of the aerospace and electronics industries. The strong ones demand, and their management counterparts generally agree, that the union should be coequal with management in all decisions related to the QWL process—its design, the plan of action, training of personnel, implementation, follow-through, and evaluation. The national General Motors-UAW QWL committee carries this idea to the point of formulating successive three-year strategies for cooperation.

In organizations with collective bargaining agreements, a joint letter of understanding is usually issued by the top official or officials of management and the union. The statement commits both parties to a joint effort. They agree, as one typical statement puts it, "that increased involvement of employees in matters affecting their work holds promise of making work more satisfying and can contribute to improving the workplace." Emphasis is placed on cooperative problem solving to achieve shared goals. The letter of understanding, distributed throughout the union and management organizations, makes clear that all QWL activities will proceed *outside of the established collective bargaining process*. The contract spelling out the rights and responsibilities of each party remains inviolate. The 1980 National Memorandum of Understanding between AT&T and the Communications Workers of America, for example, states that "QWL efforts must be viewed as a supplement to the collective bargaining process," and that "the integrity of the collective bargaining process, the contractual rights of the parties, and the workings of the grievance procedure must be upheld and maintained."

To implement a joint effort and to oversee progress, some national agreements create a joint national committee made up of the highest officials of both organizations. At Ford, for example, the committee is composed of three members appointed by the vice-president/director of the UAW-Ford Department and three members appointed by Ford's vice-president for labor relations.

Implementing Change at the Grass Roots

It is at the shop-floor or office level that the real action takes place. Here problems are solved through participation and involvement in small groups, variously called quality circles, participation teams, problem-solving groups, work teams, and so on. A QWL project can function for several years without formal groups, as at

Tarrytown, but a well-defined mechanism greatly increases the chances of survival beyond the initial enthusiasm. The tapping of creative potential and the implementation of change begin with careful preparation by managers and other leaders at the local level, in a plant, office, agency, operating unit, or other identifiable part of a larger organization.

Assuming that formal statements of commitment have been communicated from headquarters, the local orientation process may be launched, with personal visits underscoring for local management the total support of top management. The definition of "top management" for this purpose has to be determined pragmatically. What matters is that everyone at the local organization recognize these individuals as carrying enough authority to sanction any local break from tradition. In a multidivisional enterprise, a divisional president may do. On the union side, a regional or district leader might be the equivalent. Assisted by top professional personnel, the orientation includes a discussion of what QWL is all about and stresses the importance of flexibility and local self-determination in designing and implementing an effort. Some employers have found it helpful to bring in representatives from other organizations that have had successful experiences.

Where there has been a joint union-management declaration of support, the initial orientation sessions should be attended by representatives of the national joint union-management committee. Top union representatives may wish to meet with all union officers as a separate part of the orientation process. Union leaders and their counterparts in management consistently affirm that no QWL effort has a hope of succeeding without the interest of, and at least a modicum of commitment by, key local labor and management persons.

As a first step in organizing for QWL, a coordinating or steering committee is created. Composition and size can vary, depending on the size of the organization, but in larger industrial organizations it usually ranges between 8 and 12 members. As a rule, the committee should include key decision makers in the organization, such as top line and staff personnel. Where there is a joint union-management agreement, it is advisable to have equal representation. Each party usually selects a cochairperson; often these are the top manager and the shop chairman (or local union president). A committee made up exclusively of lower-level supervisors or staff persons may not command the respect of the organization, especially in the initial stages of the process.

The committee makes policy and plans the overall strategy for getting the effort underway. It decides where and when to begin. It determines the goals, the rough timetable for achieving them, the structural arrangements, and the personnel requirements. It draws the boundaries of its jurisdiction. In the case of a joint union-management committee, the parties reaffirm that the labor-management contract remains inviolate. Under no conditions should the QWL process serve as a substitute for the grievance procedure, nor should it abrogate the rights and responsibilities of management and union as spelled out in the contract. However, it may also be desirable to have a written understanding that work units may apply to the negotiating parties for permission to experiment with temporary deviations from the contract. As it considers plans for setting up quality circles or other kinds of participation groups at the grass-roots level, the committee should make a public declaration that the purpose of QWL is not to increase the work pace or work loads of individuals, that no one will be laid off as a result of QWL, and that whoever chooses to become involved does so voluntarily (at least with respect to nonexempt personnel).

As soon as the union decides to take part in the work-innovation process, the parties should issue a joint statement which outlines (1) their objectives, and (2) specific safeguards for employees.

RECOMMENDATION ▷▷▷▷▷▷▷▷▷▷▷ 28

The *manner* in which the committee communicates its reason for being is crucially important. Its central message is that it exists to help all members of the organization become more involved in decisions affecting their work lives. This includes the improvement of relations at all levels as well as improvements in operations. The creation of the committee should be announced through every available medium, such as plant newspapers, newsletters, bulletin boards, regularly scheduled meetings, and the like. Some organizations also schedule a series of employee orientation meetings at which key members of the committee provide information and answer questions.

There are recognized dangers in saturating an organization with information. It can raise expectations of immediate action, with immediate results. All pronouncements should reiterate that it will be many months before sizable numbers of people become directly involved in the participative process. Cynics in supervision will not at first catch the significance of QWL as a new style for running the organization. Some will view it as "just one more fancy program that management is forcing on us." Hard liners in the union will suspect that QWL is just another "management gimmick" to get workers to work harder and faster. Front-line supervisors may worry that their authority will be undermined.

These are dangers. But there are no alternatives. To launch the program surreptitiously, or to give the impression that QWL is an "experiment," will raise even more questions throughout the organization. It should be made clear that although the initial intervention—the establishment of a few quality circles or participation groups—is limited in scope, all members of the organization will be kept informed of progress as the effort expands.

Apart from what the committee does to communicate its purpose, the actions of top management are important in preparing and stimulating the management organization. Too often the impression is given that QWL is something of benefit to rank-and-file personnel only. But "managers are people." As managers, they have the responsibility for carrying out company policy; but as employees, *their* desires for a healthy quality of work life are no less intense than those of the people they supervise. Supervisors must be persuaded that they too will have genuine opportunity to benefit from the process through more opportunities for involvement and increased self-determination as to how they will structure their own jobs and relationships.

Thus, the total environment must be conditioned and prepared so that the transplant of employee involvement (with or without a union) into the culture will "take" and not be ultimately rejected. Understanding and acceptance on the part of the management group are essential. Before employee participation groups are formed, top management must take time to interact with lower-level managers and supervisors. Something more than issuing memos of intent is required. In successful cases the CEO has personally met with middle and lower-level management to underscore the announcements of a coordinating committee. Also, in such cases, top management has explained that the regular system of performance evaluation will henceforth include judgments of managers' success in supporting the QWL changes.

RECOMMENDATION

29 ◁◁◁◁◁◁◁◁◁◁◁◁◁

The parties should make sure that everyone concerned understands from the beginning that, even though the process starts in one or a few select areas, the entire organization will ultimately be involved.

Coordinating Initial Steps

The ground-level activities of the QWL process will quickly bog down in details unless someone is designated to coordinate (or facilitate) them. The choice of such a person is usually left to management. In some instances the joint committee chooses, or has a veto. Where there is a union, the ideal solution is to appoint two coordinators, one from each side.

What kind of background should a coordinator have? It is common to select someone who has had professional experience in working with groups—an OD facilitator, for example. Certainly it should be a person who has the respect of line and staff managers and of nonsupervisory employees. In many instances a well-respected person with no formal training in group dynamics or problem solving is designated. It is advisable that such a person quickly undergo training either through courses given within or outside the organization, or from the consultant hired by the organization.

To whom should the program coordinator or facilitator be responsible? Common practice has it that the person reports to the local manager of personnel or human resources. However, where management gives the QWL effort high priority, the coordinator reports to the top line manager of the unit. Access to the top is essential in the launching period. Where there are two coordinators, one from management and one from the union, they "report" to the joint union-management coordinating committee. In one case—the national QWL agreement between the Communications Workers of America, the International Brotherhood of Electrical Workers, the Telecommunications International Union, and AT&T—union coordinators are paid out of union funds, except for activities that are truly joint in nature.

Forming Participation Teams

After communicating goals and purposes and staffing the QWL effort, the next step is establishing participation teams at the grass-roots level of the organization. When a local union and management have had a rocky relationship, as at many construction sites, they may begin by having a single group, such as a "labor-management committee," for the entire site. Such a group, although useful, tends to focus on site-wide problems and does not give most employees an opportunity to participate directly.

Participation teams go by a variety of names, but the most popular is quality control circles, or simply quality circles. Derived from Japanese example, a QC circle is a small problem-solving group composed of six to twelve workers, usually under the same supervisor, in a particular work unit. In the "pure" or original form of QC circles, participants are taught analytical techniques for solving job-related quality problems. In industrial operations, a group's efforts may include reducing defects, scrap, and rework; improving tooling and accessory production functions; or even, as at Ford's Sharonville, Ohio, plant, helping to plan and execute a major change in a production line. These are expected to result not only in lower costs and increased productivity, but also in the improvement of skills, teamwork, leadership abilities, morale, motivation, and sense of achievement. At the core is the recognition that rank-and-file employees can make an important contribution to the organization and to the quality of their own work lives.

Although these are the basic purposes and functions of the core participation team, the Japanese, as well as Americans, have changed its function and scope. What are known as *jishu kanri* activities in the Japanese steel industry are an advanced form of QC circles, whose scope extends beyond quality control and cost savings into a wide variety of functions to improve organizational effectiveness. Similar problem-solving groups exist in American industry, except that in union-organized plants contractual issues sometimes limit the subject matter that work groups can deal with.

Groups vary in size, composition, and structure. The most common form, as in Japanese QC circles, is the group made up of a supervisor, who also acts as leader, and a group of the supervisor's own employees. In some organizations the groups elect their own leader or use a trained facilitator. In many places the group consists of five or six hourly operators plus representatives of the management support groups that service a particular production unit, such as process engineering, maintenance, inspection, and cost accounting. The advantage of cross-functional representation, a common format at Ford, General Motors, and Honeywell, is that the expertise needed for the diagnosis and solution of problems is always at hand. Alternatively, a group made up of nonsupervisory people is given the authority to call upon expert advice as needed.

Most successful programs start on a small scale with three or four groups—no more—to begin with. One compelling reason is that a single program coordinator or facilitator is limited in the number of groups he or she can introduce to, and train in, problem-solving techniques. Also, valuable experience can be gained from the "pilot" groups, enabling the program to be modified or fine-tuned as it progresses. Even if enough trained facilitators are available to launch a broad-scale effort under a "master plan," they run the risk of being "locked in," with little room for maneuvering.

In which units of the organization should the experimental work groups be set up? Many organizations have made the mistake of choosing units where the most serious operational or managerial problems exist, on the theory that there is more room for improvement, more problems to work on. During the start-up period, success matters more than dramatic results. It is far better to set up a participative team in a unit which is functioning reasonably well and where the climate favors innovation. The manager or supervisor should be a person who, perhaps as a result of the organization-wide communications efforts described earlier, is excited about trying something new and who believes in participation. Preferably, the manager should be both secure and ambitious. In a unionized location, it is wise to select a work unit in which the union representative is popular as a leader, interested, and willing. If a group is formed where these conditions do not exist, failure is very likely. Word of failure can spread quickly and cripple future efforts to expand the program.

The parties should launch grass-roots action in a small number of work units, using them as pilot projects in which procedures and problems can be sorted out before expansion begins. Units should be selected in which the chances of success are high and where strong leaders are in place.

RECOMMENDATION ▷▷▷▷▷▷▷▷▷▷▷▷ 30

Orientation and Training

Once a group of volunteers has been selected, the process of orientation and training begins. Few groups can plunge right in and start to function effectively. The members may have volunteered out of interest and curiosity, important ingredients in themselves, and yet not know what they are supposed to do or how to manage themselves. The first meeting sets the climate. Poorly handled, this meeting can sow the seeds of discouragement and dissension.

Usually the early meetings (which should be held on company time) are led by a trained professional, such as the program coordinator or facilitator, working under the direction of the principal coordinator or the consultant. A high-level manager, perhaps a member of the top policy or coordinating committee, will give a brief presentation reiterating top management's commitment. If the effort is cosponsored by the union, it is wise to have a local officer also underscore the commitment of the national and local union. If three or four groups are starting up in the launching period, one method of stimulating interest is to bring in a successful group from some other plant or office in the corporation. To have others at the new groups' level share experiences with them can be as convincing as any other approach.

The groups will need a set of ground rules. The primary rule is that the group will establish its own ground rules. It will choose its own leaders (unless the structure agreed upon by the coordinating committee calls for a quality circle to be headed by its supervisor). It will decide what problems to work on, what help it may need, and where such help is available.

Not too much should be loaded on a group at its first meeting. If the organization has been run in the traditional bureaucratic manner, the members need time to think about the new message they are hearing, which says, in effect, that management is willing to change to a more participative style. They need time to accept the fact that their ideas for improving operations are going to be listened to—and acted upon. And they need to think about the challenge, which is that their assignment is not to identify a lot of problems and "dump" them on management, but rather to present workable solutions.

Commonly, groups are exposed to several sessions of group training exercises *before* they get into identifying and solving workplace problems. There is no rigid format for such "up-front" training. Much of it has borrowed heavily from organizational development, dealing with motivation, interpersonal relationships, and group dynamics. Experiential exercises or cases are used to illustrate concepts drawn from Blake, McGregor, Herzberg, Argyris, and other behavioral scientists. Role playing is often used to sensitize the group members to their own behavior as communicators and listeners. Simple behavioral tests stimulate feedback and discussion. Brainstorming is a popular technique, especially in the initial meetings. Nondirective techniques, in the tradition of T-group sensitivity training, are sometimes used.

Cognitive training may include analytical problem solving, such as Kepner-Tregoe, "fishbone" cause-and-effect analysis, Pareto analysis, median charting, and statistical quality control.

Serious questions have been raised about the value of giving this kind of training "up front". There is little evidence to prove that over the long run it enhances a group's ability to solve practical operating problems. What goes on in the classroom does not necessarily transfer to the workplace. The experiences may be intellectually stimulating, but whether they become internalized, whether they actually change behavior, is far from certain. Also, when conducted by paid consultants, they can be expensive, especially if there is to be broad-scale expansion of quality circles and problem-solving groups throughout an industrial plant or large business office. Over one and one-half million dollars was spent in one large American plant on three days of such training for all hourly employees, yet there was little systematic follow-through on the shop floor in problem-solving groups.

An interweaving of formal training and actual problem solving appears to be preferable to separation in time. In a number of start-up efforts, the group members, after initial orientation about QWL and their new role as problem solvers, are encouraged to plunge right in on issues that directly concern them. From the first day they are asked to list problems *they* want to talk about. They assign priorities and begin at once to analyze one or two problems of greatest concern. A few sessions may proceed in a desultory way, until the group begins to see that it needs help. At that point, the skilled facilitator or group leader introduces them to formal problem-solving techniques or behavioral concepts. Now, however, the substance of what they are dealing with is a specific shop problem they have identified and want to solve. The training is no longer just an interesting intellectual exercise, but is folded into the problem-solving process itself.

In time the groups learn to manage without the help of a trained facilitator. The group's own elected leader takes over and keeps the meetings on track. The leader sees to it that written records are kept and defines each problem the group has chosen to address. Information on the diagnosis of a given problem, the causes, and alternative solutions is recorded. Assignments are made, preferably voluntarily, to a group member or a subgroup to contact others in the organization who can provide technical assistance. The group may even invite support personnel to join the group to discuss the problem and offer advice on alternative solutions. The leader or group secretary sees to it that minutes and worksheets are distributed to the group members in preparation for the next meeting. Management provides typing assistance.

The first few months are ticklish. Groups wish to see action soon. They are testing management: they want to know if their efforts are really causing changes or whether they are engaged in a pseudo-participative exercise. Many smaller problems can be solved by the group members themselves merely by explaining a problem and its solution to a supervisor or other appropriate person on the management staff. On more important problems, which require substantial systems changes or capital expenditures, the groups must be prepared to present a full-scale analysis and rationale for their recommendations. Again, they may wish to call in the QWL coordinator for help in developing presentation techniques.

In any case, feedback on recommendations must be prompt in order to maintain interest and enthusiasm. When a proposal is turned down, the members of the group need to know why—and they will usually accept an honest explanation. Many QWL efforts in American organizations have died on the vine because of poor feedback and lack of action follow-through. These signal that management has only paid lip service to the idea of participation and that other problems are considered more pressing at the moment. Employees are quick to sense this, and, if they do, soon lose interest in trying to bring about change.

RECOMMENDATION
31 ◁◁◁◁◁◁◁◁◁◁◁◁◁◁

Management should deal with work-group recommendations in a timely, respectful, and open-minded fashion. When modification or rejection is called for, management should state its reasons clearly and offer to help if the group wishes.

The Support Role of Supervision

Many quality circles or problem-solving groups are launched without full recognition of the role of supervision in the process. Too often supervisors feel left out, unless they happen to be members of a group. In spite of all the communications from top management urging everyone to cooperate in the new effort, front-line supervisors have natural concerns about the integrity of their authority. A supervisor, hearing that the problem-solving groups are free to tackle any issue they choose, may well feel confused and threatened. Many programs have started up with much enthusiasm by the work group, only to flounder because supervisors are turned off.

In the more successful programs, supervisors attend workshops and are exposed to all of the orientation, training, and information materials *before* the problem-solving groups are formed. At the very least, supervisors in whose departments groups are being organized should receive full exposure. The principle of involvement should apply to them too, and they should participate in the planning process.

Once the groups have started up, the supervisors must be kept informed about what the groups are doing. Many of the operating problems being studied by the groups are of direct concern to the supervisor. He or she is often the first person to be contacted by group members in collecting information about a problem. If a supervisor's boss is not committed to QWL, the supervisor is likely to ignore requests from the group members in the old political game of knowing where one's bread is buttered. But when the climate of participation is favorable throughout managerial ranks, the front-line supervisor will see that cooperating with the groups is important. In some organizations, periodic performance reviews include an evaluation of a supervisor's support for, and help to, the work groups. This formal control mechanism helps, but it does not in itself assure cooperation. What ultimately brings supervisors on board is the perception that the groups are, in fact, helping them solve problems which would otherwise generate crises and emergencies. A volunteer group cannot be ordered to take up a problem that is important to the supervisor, but the supervisor can arouse its interest by candidly discussing operating priorities and making clear that the group's advice will be appreciated. In the most successful cases in industry, this mutually supportive relationship of supervisors and QWL groups has gone far to eliminate supervisors' feelings that their authority has been threatened.

Final Note on Implementation

The lessons learned from the initial pilot groups are invaluable for the next phase, "diffusion" to other parts of the organization. The coordinating committee has real-life data on what works and what doesn't work. It is in a position to think about the basic structure of the problem-solving groups, about the format of training and orientation, and about the issues that have gotten in the way of progress or that have facilitated the process. It can assess the response of supervisors and of support personnel. It can discuss collective bargaining issues that may have arisen as a result of the experience to date, especially the reaction of shop stewards. The committee can advise the operating head of the plant or office of changes in policies or procedures which may have to be instituted.

7.
Growth and Diffusion of Work Innovations

Typically, pilot projects, or trial sites, are chosen and nurtured with the end in view of growth and diffusion of the work innovation—first to similar or related activities in the same location, and later to other geographical locations. This strategy may be assumed to be a less difficult stage of development than the initial effort, but it has pitfalls.

Initial success in one or several units does not necessarily mean that the process can be transferred easily to other units. To begin with, a work innovation that has been successful at one site may not lend itself to imitation. Furthermore, obstacles due to the personalities involved, organizational differences, and corporate and union politics are always present.

A major roadblock to growth and diffusion is the reaction of other organizational units to the success of the pilot project. Although the productivity gains of a highly touted pilot project may cheer its originators, the very success of the program—and of the unit which initiated it—may be a threat to other units. The innovators cannot help but attract a great deal of attention and outshine the traditional organizations, whose people look like losers by comparison. The natural response of nonparticipant managers, then, is to question, doubt, criticize, and challenge the innovations of work units with which they compete for attention and recognition. In many cases, they may feel that it is easier and safer to kill the "experiment" than face the painful task of playing under a new set of rules.

A Strategy for Diffusion

It is necessary, therefore, to prepare carefully for the growth and diffusion of work innovations beyond the initial pilot program and to provide a maximum of assistance. Developing and implementing a strategy for diffusion will take as much effort as preparing the organization for the initial launch, and more so if work innovations are to be extended throughout a multilocation enterprise.

There is no universal formula for the process of diffusing work innovations. Every organization must tailor the process to suit its own history, culture, and technology, learning from practical experience how to do it.

Those who champion the program will want to expand it as rapidly as possible. Thus, line managers often feel pressed to show quick results. It takes courage to resist such pressures. On the other hand, if diffusion is too slow, the organization will become discouraged and expectation will be dampened. One thing is certain: the diffusion process takes a great deal of time and energy and requires those at corporate level to exercise considerable patience even while they transmit strong signals of support.

Another aspect of commitment should be considered before moving ahead: the recognition that worker participation, by definition, implies a shift of power. If the initial small pilot efforts have been successful, it is because the employees have come to realize that they now enjoy greater freedom to take part in decisions which affect not only their immediate jobs but the overall performance of the work unit. They realize that they have gained the power to change things, but until they can see benefits for themselves, they will resist the actual exercise of power. That power goes beyond the right of employees merely to meet once a week and choose problems they wish to work on. It includes the right of access to information formerly shared only by managers and supervisors. Employees have learned how to lobby for change in a way that goes beyond the usual employee-management relationship. Where there is a union, this lobbying process adds another dimension, outside of the regular bargaining relationships. To members of management, this subtle shift may appear to be a loss rather than a gain. Until they can see benefits for themselves, they too will resist participation in the pilot program.

That is not to say that the basic prerogatives of either management or the unions will be undermined. Rather, the decision-making process changes so that fewer problems are resolved unilaterally by management. If the pilot efforts were well conceived and carried out, managers will see that they contribute to the effectiveness of day-to-day operations, and union representatives will recognize that they reduce the number of unsolved operating and work-flow problems that irritate employees and eventually blow up into grievances.

Planning Diffusion and Learning from Experience
The first steps in expanding and diffusing work innovations within a single location grow from the direct and immediate experience of the pilot groups. The top coordinating committee works out a strategy for diffusion, without necessarily creating a complete master plan. Progress begins with small tactical advances. The coordinators may not know what the whole "forest" looks like, but they can see a few "trees" and discern some paths to follow.

The coordinator of a successful quality-circle program in one division of the Martin Marietta Corporation has described the period of transition, from original pilot experiments to full-blown program, graphically:

> In short, we could progress only by engaging in tactical advances, with the faith that we were moving in the right direction. We could see trees, one by one, but we could not see the forest or the path through it.
>
> Not until after a year of experience with quality circles, actually solving problems and interacting with the wider organization, did we begin to see the forest—the broader picture. We came to see a quality circle not as a "thing" but as a management "process" in which the entire organization—including people outside the circles— participate, and must participate, for the program to succeed. Our subsequent efforts were aimed at developing systems to support this process.
>
> Since recognizing quality circles as a management process, our division has expanded further. We recognize a "countryside" beyond the "forest," relevant to the success of the quality-circle process. Specifically, the success of our quality-circle process is affected by the kind of infrastructural support it can receive from outside the organization.[1]

This linkage between the "trees" (pilot groups) and the "forest" (the site) and the "countryside" (the rest of the enterprise and outside organizations) is crucial. In a well-known case of failure, diffusion was aborted because managers, union representatives, and others did not understand what the experiment itself was all about and were unprepared to take the next steps to broaden the effort.

As soon as the pilot programs have been recognized as successful, the parties at the local level should draw up a strategy for diffusion to the rest of the site, based on the early experiences and tailored to the unique conditions of the organization.

RECOMMENDATION ▷▷▷▷▷▷▷▷▷▷▷▷▷ 32

The strategy for diffusion should be adjusted for any weaknesses or problems brought to light by the pilot programs. A number of questions need to be considered by those who will be responsible for coordinating the expansion and growth of the programs:

—Have the pilot projects stimulated a lasting commitment on the part of key members of the organization?

—Is there so much "turbulence" in the system (due to economic conditions and work-force fluctuations) that expansion becomes risky?

—Has there been so much turnover among key personnel that commitment is now weak?

—Are the needed financial resources and people available to support the expansion?

—Do employees feel that their jobs are relatively secure?

—Have the initial participative teams maintained a high level of interest in solving problems?

—Did the initial groups learn not to use the problem-solving sessions as platforms for airing complaints?

—Did the line and staff people, when called upon by the groups, provide adequate support?

—How many new ideas did the groups generate?

—How many solutions generated by the initial groups were implemented?

—How effective were orientation and training for the problem-solving groups and other key persons in the organization?

—Do front-line supervisors in the pilot areas feel that their own jobs and operations benefited from the work done by the groups?

Negative responses to these questions do not necessarily mean that the efforts at diffusion should not proceed, but that effective countermeasures must be devised.

Starting the Diffusion Process

Those responsible for getting the program to expand need information on which to base a decision as to when and where to start. Members of the coordinating committee can pick up clues through the grapevine or other informal communications. The coordinator or facilitator, knowing from experience what has worked, will have some ideas; so will members of the pilot groups from their own contacts.

Many organizations use surveys as information vehicles in making the transition from experiment to full-scale diffusion. Covering hourly and clerical workers as well as supervisors, a survey can point the planners toward people and departments that are likely candidates. Useful questions for the survey include:

—How much knowledge do people at the location (but not directly involved) have about the employee-involvement program?

—What perceptions do they have with respect to its purpose and goals?

—Where is the potential leadership and commitment? In short, who is ready to move ahead?

Many experienced practitioners of QWL advocate "loading the dice"—that is, not leaving the program to chance. They want assurance ahead of time that there is a built-in readiness and interest. It is worth repeating that the attitude, style, and behavior of top management have much to do with providing a favorable climate. A kind of "healthy infection principle" operates in the spread of QWL efforts. People hear what is happening, see that participants like it, sense that top management supports it, and decide that they want to be involved too.

In forming new groups, the basic principle of voluntarism still holds. No one should be forced to join. Nor should selection be entirely in the hands of the coordinating committee. Middle managers, in particular, tend to be skeptical about the entire effort if they feel left out. "Participation" is the fundamental goal—therefore this should be the goal at every stage of planning.

The number of groups is limited by the number of people who can do the orientation and training. There may be general enthusiasm to join groups, but not everyone can be accommodated at once. The reasons for not moving ahead full throttle should be openly and frankly discussed, and employees should be assured that they will be welcomed as soon as possible.

Many employers have underestimated the costs of diffusion and left individual locations to bear the costs on their own. In successful cases, divisional or corporate management lays out strategic plans for underwriting the diffusion process, at least in the early stages. Usually a plant or office needs a professional consultant (either internal or external), with experience in other units, to stimulate action.

One coordinator or facilitator may be able to coordinate the activities of 10 or 12 groups at a time. But at a certain point additional trained personnel may be required. Some locations may be able to underwrite the costs. Others may have to draw in trained staff members from divisional or corporate headquarters to assist them.

Eventually the plant or office must be self-sufficient in bearing the costs of diffusion. For this purpose a permanent in-house cadre of trained leaders, coordinators, and facilitators is built up. At Ford's Sharonville plant, people with line operating experience were assigned to each of eight production superintendents as "mini-coordinators" for their departments. Each person had completed training and orientation with the plant coordinator/trainer and had had actual experience as a member of one of the initial problem-solving groups. Each person was responsible for starting up new groups in a particular department and worked within the guidelines of the plant-wide joint union-management coordinating committee.[2]

In diffusing work innovations within a location, the parties should direct their efforts toward work units whose supervisors and employees are most ready and willing to take part. However, they should make clear from the start that all work units at the location will eventually have the opportunity to participate.

RECOMMENDATION ▷▷▷▷▷▷▷▷▷▷▷▷ 33

Diffusing Change—How It Is Structured

As the groups proliferate, a sort of parallel or alternative organization emerges. Superimposed on the regular hierarchy is what someone has called a "heterarchy." New structures, with new roles for employees and supervisors, are put into place, supplementing but not necessarily replacing the old ones. New communications links, for sharing knowledge about how the organization can function more effectively, open up.

As an example of how this new heterarchy might look on an organization chart, the regular organization may have five operating or service departments. Each department comprises 200 employees, divided into five sections of 40, each section headed by a supervisor. The goal is to have one quality circle in each section. Each circle has ten members, sometimes including persons from the support staff.

At the top of the structure is a coordinating committee of high-level managers and union representatives (where there is a union). Reporting to the committee are two full-time coordinators, one drawn from management and one from the hourly ranks. The coordinators, who are trained in all aspects of QWL and were involved in setting up the pilot groups, will manage the diffusion process.

Next, a minicoordinator is appointed and trained to serve each department and its five problem-solving groups. At Ford's Sharonville plant, such a person serves in two capacities—as a planning assistant for the department manager and as coordinator for QWL activities. He or she is the communications link between the regular management hierarchy and the emerging heterarchy of the work groups. The minicoordinator ensures that line management and the top coordinating committee are kept informed about the progress of the groups. Where groups are having difficulty in getting their recommendations carried out, the department coordinator alerts the appropriate action takers in line and staff management. But coordinators must not make operating decisions; these must be worked out between the groups and management.[3]

In addition, the departmental coordinators provide orientation and training for the new groups as they are formed, train the group leaders, and nurture the groups to the point where they are on their own and functioning with a high degree of autonomy.

There are other possible forms for structuring the diffusion process. In some organizations all group coordinating and training activities are carried on by full-time trained professional facilitators. The advantage of full-time facilitators is that when problems arise, they can react immediately. They can quickly spread ideas and insights from work group to work group. And since they do not report directly to the line supervisor, they are less subject to conflicting orders and political pressures. Once a department's work groups are more or less on their own, the coordinator can be freed to help establish new work groups and structures in other parts of the organization.

A quality circle or problem-solving group contains only a fraction of a work section or department, but it does not function in isolation from the majority of employees. When a group is running well, its members are, in effect, representatives for their entire work section. They do not simply huddle together once a week, identify problems, and recommend solutions. They keep their workmates on the shop floor or the office informed about the problems which the group is tackling. In successful programs they solicit ideas and suggestions from their workmates and bring them into the group discussions. Group leaders or facilitators should be sensitive to this linkage phenomenon and make sure that people in the organization not directly involved in problem-solving groups are kept in contact with them.

Another way to prevent isolation is to make the composition of the groups flexible. Leadership may rotate. The kinds of problems a given group addresses may change over time. New people with a special interest in, and experience with, a problem can be brought in. Members of the group may drop out when the problems cease to hold their interest, and sometimes the entire group goes out of existence. Over the course of a year, it is not unusual for all employees in a section of 30 to 40 people to have spent some time in a particular quality circle.

As the diffusion process continues, it has been found useful to "seed" new groups with people who have previously been members of successful ongoing groups so that they can share their enthusiasm and experience with new members. They make excellent "salesmen." As peers, they can bring convincing evidence to show how and why the involvement process really works. The enthusiasm they transmit to a new group can be more effective than the formal orientation provided by facilitators and coordinators.

Sometimes the members of a new group attend a session of a group that has been operating for some time. Although they may not be interested in the particular problem the group is working on, they can observe the involvement process. They can sense that a well-functioning group has the power to get things done.

As a substitute for the attendance of new groups at existing groups' meetings, some organizations have made videotapes of groups meeting or presenting their recommendations to management for action. Another method is to take members of a new group to the office or shop floor to see a new system or process that has been installed on the recommendation of an experienced problem-solving group. Still another reinforcing mechanism for orienting newly formed groups is to have the plant manager or office head meet with the group in the first orientation session. The manager may personally stress commitment to the involvement philosophy and talk about the good work done by existing groups. In a joint union-management participation effort, the head of the union may appear together with the top manager.

RECOMMENDATION 34 ◁◁◁◁◁◁◁◁◁◁◁◁◁ *The parties should use all available formal and informal means to acquaint nonparticipant employees with the activities of the groups and should encourage the groups to involve other interested employees in the groups' activities.*

In the more advanced programs the diffusion process often extends beyond hourly workers or clerical workers in the lower ranks. Large industrial plants, for example, employ many office workers. As word gets around about the quality circles and problem-solving groups on the shop floor, they too "want in." If the plant operates under a labor agreement, white-collar workers are usually omitted from the bargaining unit and would not be served by a joint union-management coordinating committee. In such a case a separate coordinating committee might be formed, including representatives of departments, such as accounting, purchasing, engineering, and so on—including not only management personnel but also volunteer clerks, secretaries, and support personnel. This salaried committee performs coordinating functions similar to those of the joint union-management coordinating committee. Its members find out who in each department has an interest in forming a quality circle. Groups are organized and proceed in much the same way as the factory groups. A full-time facilitator may be assigned to help in organizing and training the new groups.

There are no limits to the diffusion process. Honeywell, for example, considers expansion part of an overall strategy for creating a participative climate throughout the organization. Quality circles have been established not only for secretarial and office workers but also for engineering teams and for representatives in domestic marketing, foreign marketing, and sales organizations. In a few companies, supervisors have formed their own quality circles to deal with organizational problems of particular interest to them.

Other forms of quality circles have stimulated the diffusion process. The Denver Aerospace/Michoud Division of the Martin Marietta Corporation has invented special structures to meet specific needs. For example, "task force circles," made up of volunteers from several different functional departments, deal with interdepartmental issues; each dissolves as soon as its special problem has been solved. A "joint circle" is made up of employees whose operations are linked to outside suppliers or vendors. When problems arise relating to the fabrication of parts, the circle members meet or make telephone contact with suppliers. If parts and materials come from within the organization, they communicate with the department responsible for supplying them, sometimes even meeting jointly with quality circles functioning in the other departments. At a still higher level, "integrated circles" are composed of volunteer representatives linking their own organization with representatives from the government auditing agency and from NASA. These groups receive regular quality circle training and tackle major management problems.[4]

Developing Support Systems for Diffusion

The process of diffusing employee-involvement systems requires constant reinforcement and feedback. As groups show progress in solving problems, there is a natural craving for recognition. Although the groups and all those directly connected with the effort know what they have accomplished, they want their contributions to be recognized publicly.

In one successful case, a variety of feedback mechanisms are used. Every new breakthrough is posted plant-wide on bulletin boards specially designated for information on the quality-of-work-life program. One section of the monthly plant newsletter tells where new groups are starting up and gives examples of the accomplishments of particular groups. Members of top management come together to witness group presentations in which proposals may involve substantial outlays of funds or major changes in operating systems.

In a multidivision corporation, the CEO and all members of the executive staff set aside a whole day for presentations by selected work teams. This direct interaction between top management and employees from the lowest level of the organization is a potent reinforcer.

These kinds of public acknowledgment stimulate the interest not only of the groups themselves, but of those who have not yet participated. A solid base for further diffusion is built. Group members, seeing that their efforts have had an impact, experience an added sense of dignity and status. The message that trickles through the grapevine is amplified by formal communications and by official recognition from the top leaders of the organization. The broad and powerful constituency that develops helps to perpetuate the system and assure its growth. Upper-level managers, middle managers, supervisors, support personnel, and the employees themselves now feel they "own" the process.

Those concerned with growth and diffusion usually ask, "As problem-solving groups proliferate, how can interest and enthusiasm be sustained?"

The answer is simple. Most organizations are constantly undergoing change: new products or services are introduced, technology is brought in, new systems are created to meet new conditions, and new problems have to be solved. The problems cannot be solved by management alone in the traditional decision-making process. Employees at the grass-roots level have the knowledge, and they *want* to be involved in solving constantly emerging problems.

RECOMMENDATION 35 ◁◁◁◁◁◁◁◁◁◁◁◁◁ *The parties should ensure that the accomplishments of problem-solving groups receive public recognition from management.*

Transferring Success outside the Immediate Organization

In large enterprises diffusion to other divisions and locations is stimulated and coordinated from the top, often starting with joint national union-management committees, which set the tone. Divisions and units are encouraged, not ordered, to start up new programs. Financial support is provided, but for the most part individual plants have latitude to proceed in their own way. Dictated rules and regulations, even when based on brilliantly successful programs, are resented at the unit levels; moreover, they fly in the face of the fundamental philosophy of QWL, which is that participation and involvement are the business of those who have to live with the decisions.

On the other hand, without support from headquarters, decentralized efforts may be unable to spark action elsewhere in the organization. In a few cases, individual plants or offices have initiated a successful program on their own, but their success has neither spread to other plants nor influenced top corporate management to espouse the idea. For example, the radical change in work systems at the Gaines pet food plant in Topeka required deviations from established corporate procedures. The corporate staffs for personnel, accounting, engineering, quality control, and so on, could not stop Topeka, but they did prevent other plants from operating in the new mode. Also, because few corporate executives had a sympathetic understanding of the Topeka plant, there were no "inside" champions of the cause.[5]

In the case of an anonymous multiplant manufacturing/ engineering firm, more than 60 experimental self-managing teams sprang up in various traditionally managed locations over a period of a decade. All were local management initiatives. Some were spontaneous in origin; others resulted from word passing along the grapevine. Many worked well for a while, but none endured very long and no real diffusion took place. Corporate managers' interest in them never reached the point of commitment or formulation of a strategy.[6]

Another company, taking a different approach, encouraged the start-up of a QWL effort in one of its many locations. The success of this experiment led top management to adopt QWL as a matter of policy and urge all divisions and plants to apply QWL principles to their own organizations. They promised financial support, and professional personnel brought plant and division managers into an orientation conference and consulted with the respective unions. They did not dictate whether or how each would proceed.

Another effective strategy is to use modest successes to spread the word on a voluntary basis. The corporate committee, or more likely a group at the division level, brings top line and staff managers together and exposes them to the activities going on where an effort has been launched. After general orientation, representatives of the pilot group describe in detail, and respond to questions about, the structure, the process, the successes, and the roadblocks which the unit has experienced. The representatives of locations where the process has not yet been started learn from their peers, and the presence of top division or corporate managers underscores the commitment from the top.

Perhaps the most effective method for promoting the "healthy infection principle" is intersite visits, in which an interested group of managers (and union representatives, where there is a joint agreement) from one site spends time observing directly how another site is managing work innovations.

General Motors has a well-integrated strategy for diffusing work innovations through a large, multidivisional enterprise, which has stood the test of a decade of practice.[7] The ingredients are: (1) a strategy for system-wide change, (2) an early decision to institutionalize change, (3) an organizational culture that encourages and supports change, (4) a variety of mechanisms to interchange experience and learning among sites, (5) a strategy for shifting responsibility and authority downward, and (6) a flexible reward system.

System-wide change began in the late 1960s with a number of different programs at different sites: team building, job redesign, training of supervisors and hourly workers, and business teams. None of these were diffused to other sites because there was not yet a strategy for doing so. As a first step in creating one, a top operating executive convened a meeting of the heads of nine major divisions; he challenged the participants to begin work innovations, and he announced that at a second meeting to be held in six months he would expect to hear what they had done in the interim. Meetings in this series have continued until today. Called the Annual Executive QWL Conferences, each is attended by over 250 top operating executives, in the United States and overseas. In 1981, GM's International QWL Conference was followed by seven one-day regional meetings, which involved 1,600 managers and 265 union leaders.

To support system-wide change, the corporation began in the early 1970s to select and train GM employees ("usually highly credible, middle-management people")[8] as OD/QWL specialists within their own plant, division, or staff organization instead of relying on external consultants. The corporation also decided to place responsibility for QWL in the *line* organization.

As a corollary, each site had the authority to determine the nature and pace of change it would undertake—subject to the following minimum standards: biennial administration of GM's standard QWL survey, creation of a local top-level committee to oversee QWL, a local statement of QWL philosophy and principles, a process to familiarize local employees with QWL principles and techniques, and qualified internal resources to assist in work innovations.

Since intersite learning is retarded when one site's success is invidiously compared with the lack of success of others, or when one site's model is forced on others, GM adopted a rule to the effect that "all projects must be seen as a joint effort on behalf of the entire organization. If comparisons are to be made, either compare before and after conditions within the same unit, or use another division or the Corporation as a frame of reference. . . .The locations that receive the greatest publicity and. . .the greatest acceptance are those regarded as learning models at the cutting edge of the QWL process."[9]

Mechanisms for intersite learning include conferences, seminars, workshops, projects, printed materials, and videotapes. In addition, there are networks of people from all parts of the corporation who meet regularly. For example, people who have overall resource responsibility for QWL activities within their divisions or staff groups meet quarterly, each time at a different in-company site so that local operating people can attend. Another network draws together managers of plants with nontraditional work systems, who also meet at a different site each time.

GM's corporate commitment to QWL spurs diffusion. Periodic reviews of QWL activities are held by GM's executive committee, public policy committee (composed of outside directors), and science advisory committee. In 1981, the full board of directors received an updated report of QWL activities. Top corporate executives take part in QWL-related off-site meetings, project reviews, conferences, and on-site visits. The group executives review with the executive committee the QWL annual survey results and activities.

Besides corporate initiatives, there are many joint union-management activities to help in diffusion. The GM-UAW national QWL committee, created in 1973, has overseen the establishment of joint committees at more than 70 locations (including a few with International Union of Electrical Workers [IUEW] and United Rubber Workers [URW] representation). Joint national committees on work attendance and on product quality have been created, and again some divisions and plants have adopted the model. Joint committees sponsor seminars, off-site meetings, and conferences; receive visitors from other locations; and help select and train QWL coordinators.

The parties at corporate level should develop a strategy for diffusing work innovations throughout the enterprise. The strategy should allow each location great latitude in determining its own form, process, and pace of change, provided that **RECOMMENDATION** ▷▷▷▷▷▷▷▷▷▷▷▷ **36** *reasonable minimum standards of progress are met. It should also provide maximum opportunity for managers to learn about work innovations in all parts of the organization, through site visits, conferences, educational programs, networks, and the like.*

Until recently, no formal arrangements have been made to reward managers explicitly for the success of work-innovation efforts. Informally, it is known that in a few large companies managers have been promoted for their QWL successes. In one of these cases, a person in a staff position who had spearheaded a QWL effort at a lower level was elevated to corporate vice-president for personnel and labor relations. In another instance, a local production manager who had stimulated employee involvement was promoted to plant manager of another, larger plant.

One problem in rewarding managers is that the QWL process often takes a long time to show results. In the American tradition a manager is evaluated in the short term, semiannually or annually, under the standard MBO or performance-appraisal system. The overriding criterion of "return on investment" in a short time period prevails. In some companies, such as Ford and General Motors, the appraisal system now formally includes an evaluation of each manager's efforts to promote QWL, and the individual is not expected to demonstrate immediate results. In another large company, the CEO publicly announced to the management organization, "We are committed to a new philosophy of greater employee involvement, and you are going to be paid for creating this kind of environment."

RECOMMENDATION 37 ◁◁◁◁◁◁◁◁◁◁◁◁◁◁

As part of the diffusion strategy, top management should conspicuously reward managers who achieve outstanding performance through work innovations.

NOTES

1. Philip C. Thompson, "Quality Circles at the Martin Marietta Corporation, Denver Aerospace/Michoud Division," in *The Innovative Organization: Productivity Programs in Action*, edited by Robert Zager and Michael P. Rosow, Pergamon Press/Work in America Institute Series (New York: Pergamon Press, 1982).

2. Robert H. Guest, "The Sharonville Story: Worker Involvement at a Ford Motor Company Plant," in *The Innovative Organization: Productivity Programs in Action*, edited by Robert Zager and Michael P. Rosow, Pergamon Press/Work in America Institute Series (New York: Pergamon Press, 1982).

3. Ibid.

4. Philip C. Thompson, "Quality Circles at the Martin Marietta Corporation, Denver Aerospace/Michoud Division," in *The Innova-*

tive Organization: Productivity Programs in Action, edited by Robert Zager and Michael P. Rosow, Pergamon Press/Work in America Institute Series (New York: Pergamon Press, 1982).

5. Richard E. Walton, "The Topeka Work System: Optimistic Visions, Pessimistic Hypotheses, and Reality," in *The Innovative Organization: Productivity Programs in Action*, edited by Robert Zager and Michael P. Rosow, Pergamon Press/Work in America Institute Series (New York: Pergamon Press, 1982).

6. Eric Trist and Charles Dwyer, "The Limits of Laissez Faire as a Sociotechnical Change Strategy," in *The Innovative Organization: Productivity Programs in Action*, edited by Robert Zager and Michael P. Rosow, Pergamon Press/Work in America Institute Series (New York: Pergamon Press, 1982).

7. D. L. Landen and Howard C. Carlson, "Strategies for Diffusing, Evolving, and Institutionalizing Quality of Work Life at General Motors," in *The Innovative Organization: Productivity Programs in Action*, edited by Robert Zager and Michael P. Rosow, Pergamon Press/Work in America Institute Series (New York: Pergamon Press, 1982).

8. Ibid.

9. Ibid.

8.
Institutionalizing Change

In the first half of the 1970s a wave of enthusiasm greeted the announcement of several daring experiments in work innovation which seemed to offer hope of a new culture for the workplace. This was followed by a trough of discouragement and cynicism as one project after another appeared to wither and die. What had gone wrong? Were work innovations inevitably dependent on a charismatic leader? A special culture? Fortuitous conditions?

A few investigators have taken a careful look at the alleged corpses. Their findings have laid the basis for a new understanding of what is required in order to institutionalize work innovations— that is, to make them a permanent way of life for an organization. Best of all, they discovered that some supposed "failures" were alive and well, such as the work system at the Gaines pet food plant in Topeka, Kansas, and that others had produced significant though short-lived benefits.[1]

The key finding is that preparations for institutionalization can be and have to be built into the QWL process from the beginning. Most of the necessary actions will be taken after the process is well under way, but the strategy and planning should begin at once.

RECOMMENDATION
38 ◁◁◁◁◁◁◁◁◁◁◁◁◁

The employer (and union) should include plans for institutionalization as an integral part of the strategy for introducing work innovations.

How does one determine, in the first instance, that a work innovation has, or has not, become institutionalized? In practical terms, one looks for evidence that desired changes of behavior have been incorporated into the daily functioning of the organization. An innovation is not a "thing" but a new *way* of doing things; mechanisms are established to make sure that things are done in the new way.

A Concept of Reasonable Results

As a starting place, the participants should reach an understanding of what objectives the organization is trying to attain through work innovation and what permanent changes in behavior they want to effect. The decision makers' expectations determine whether changed behaviors will persist, for two reasons. First, if expectations are unrealistically high, failure to reach them is preordained and useful achievements will be junked while seeking results that are unattainable. Second, unless there is a *common* understanding among the participants, the project managers, and the various levels of executives, there will be considerable disagreement as to whether expectations are or are not being met simply because the expectations are not shared.

Success in institutionalizing new behaviors is not an all-or-nothing proposition. Changes in behavior may vary from time to time in their persistence, in terms of the number of people in the organization involved in them, and in the degree to which they are dispersed through the organization. When the literature on workplace innovation discusses change as a success or failure, it clouds the crucial issue. Careful studies have shown that change is better described in terms of *degrees* and *levels*. Starting with that perspective can reduce the risks of investing time and energy for inadequate returns.

The Measurement Snare

After reasonable objectives have been set in terms of degrees and levels of persistence, it is necessary to monitor progress and to measure results wherever practical. Here, too, the rule of reason should prevail. Although a tracking system is needed, measurement can easily become a snare.

For example, the choice of measures is sometimes allowed to determine the nature of the innovations, rather than the other way around. Evaluation designs are frequently built into experiments, whether in private or public programs. That makes sense. However, too often those who design the measures try to impose their judgment of what should be measured on the design of the experiment itself. If the real purpose of setting up quality circles is to improve morale and product quality, it makes little sense to focus measurement on cost savings. This can restrain the scope of the activity or alter it for the convenience of the measures. The measurement should adjust to the program, not the program to the measurement.

A related danger is to limit evaluation and tracking to that which is readily quantifiable. Some evidences of persistence of desired behaviors can be quantified, some have to be judged *qualitatively*. If these are not identified, and if reporting systems are limited to pure statistics, management will not have the full picture of how the innovations are persisting. For example, numbers may tell that 50 quality circles have been created, but they do not tell how enthusiastically members participate nor how well they perform. At Honeywell, management assesses the effectiveness of quality circles on the basis of whether they (1) meet regularly, (2) set reasonable goals, (3) achieve a reasonable percentage of goals, (4) are cost-effective, and (5) interact well with management. Martin-Marietta's Michoud Division takes a different tack; it measures not only the cost savings that result directly from the circles' effort, but also the circles' effect on the organization as a whole, as revealed by measures such as those of defects, productivity, attrition, absenteeism, scrap, accidents, grievances, and employee attitudes.

In early work innovations, measurements did sometimes get in the way of results. Some projects were simply overmeasured— "pulling up the carrots to see if they are growing." It requires balance to measure what is happening without the measurements themselves intruding into the project.

RECOMMENDATION 39 ◁◁◁◁◁◁◁◁◁◁◁◁◁

The parties should measure results on the basis of what they really want to achieve, track qualitative results as well as quantitative, and avoid the temptation to overmeasure.

Following What Is Happening

Tracking the results of work innovations is important, but it is equally important to keep continual track of how effectively the work innovation is being incorporated into the daily life of the organization. Five key factors should be monitored:

1. *Knowledge of the changed behaviors.* A starting point is to learn how many people in the organization actually know what the changed behaviors are, and how much they know. Do they

know that they are supposed to behave differently? Do they know the purposes to be achieved by doing things differently? For example, while "team meetings" figure in many quality-of-working-life programs, there are cases in which people know only that they are supposed to have meetings, but do not know what they are supposed to do in them. In such a case, institutionalization has not proceeded very far and the need for better information is indicated.

2. *Performance of the changed behaviors.* Are the teams meeting? Are they meeting as frequently as they are supposed to? What is actually happening at the meetings? The idea is not merely to count the number of persons involved and the frequency of occurrence, but to compare numbers and frequency to the levels required by the change program. Only on this basis can reasonable comparisons be made.

3. *Preferences for the changed behaviors.* How much do employees and managers like or dislike the new behaviors? Studies have found that in well-institutionalized change programs, most participants like what they are doing. In change programs on the decline, negative feelings are generally expressed.

4. *Development of consensus.* How aware are individuals that other people in the organization are performing new roles or behaving in new ways? How aware are they that other members of the organization feel they *should* be performing in these new ways? The more that people believe other people both perform, and feel it is right to perform, in the new ways, the more the program is institutionalized.

5. *Development of new values.* The final index of institutionalization is the extent to which people have adopted the values that are built into the change program. Many work innovations, for example, set out to expand employees' freedom and responsibility, as in the case of semiautonomous work groups. The more these values of freedom and responsibility are developed, the greater is the degree to which the change program is embedded in the organization.

These five measures of success in institutionalizing change are steps in the process of institutionalization and usually occur in the order presented. First, people develop beliefs about the behaviors and begin to perform them. Then they develop feelings about the behaviors, and others come to be aware of these feelings. Finally, values start to evolve and spread concerning the appropriateness and desirability of the behaviors. The further this sequence has progressed, the more the change program has become institutionalized.

New Wine in Old Bottles

The difficulty of institutionalizing a work innovation depends in some measure on whether it is introduced into a green-field site or into a site currently operating on traditional lines. In either case the parties are trying to create a nontraditional culture in the workplace. In a green-field site, the new culture can be constructed almost from scratch: the employees as individuals may have to unlearn past habits, but the organization does not. That is one reason why self-managing work teams succeed more often when they are part of the original design of the workplace than when they are introduced into an established one.

Institutionalizing work innovations in a preexisting site is like grafting new branches onto an old tree. A new structure will evolve, combining the old and the new, but substantial parts of the existing structure will remain in place. In fact, the new is defined in the minds of employees by contrast to, and in comparison with, the old.

Insofar as the old structure provides a frame of reference, and to the extent that employees find comfort and security in that frame of reference, they will not want it "yanked out from under them." One important example, discussed in some detail earlier, is the blending of new management-employee relationships with established collective bargaining structures; assurance must be provided that the new approaches do not undermine the role of the union or other employee representation arrangements.

Another example is the blending of new arrangements—for example, semiautonomous work teams—with the existing hierarchy, so that supervisors gradually give up parts of their old jobs and add responsibilities as functional specialists. At General Motors and elsewhere these are referred to as "parallel structures."

Whatever the nature of the change, chances for survival are improved if the changes are congruent with existing organizational practices. For example, if supervisors are in the habit of holding weekly information meetings with employees, it is easier to shape those meetings into a consultative device than it is to start an all-new program of consultative meetings.

RECOMMENDATION 40 ◁◁◁◁◁◁◁◁◁◁◁◁◁ *When introducing work innovations at a traditionally managed site, the parties should strive for congruence, splicing the change with the existing structure, rather than making a superficial graft that won't last.*

Training

Training of employees in the new work behaviors is critical. It is also easily overlooked, for in a large majority of organizations, the adequacy of training does not stand high in the priorities of senior executives. Partly, this is because of reliance on on-the-job training as the primary means of transmitting work skills. Where work innovations have been successfully implemented, it will be found that there has been extensive initial training in the new behaviors. In order to ensure institutionalization, however, it is necessary to provide retraining as time goes on and to train new participants as turnover occurs. Membership in an organization continually changes.

In a program designed to change the practices of high-level managers in the sales division of a manufacturing firm, a retraining effort six months after start-up strengthened the program. In another study of two management-by-objectives programs in large manufacturing firms, the firm with a retraining exercise showed a greater degree of institutionalization.[2] In the case of an underground coal mine, a decrease in frequency of training after the first year contributed to the decline of work innovations.[3]

There will always be some turnover among participants in work innovations. In plants such as GM's Tarrytown and Ford's Sharonville, turnover due to the economic roller coaster has been on a huge scale. The effect has been to derail QWL programs until the new participants receive training in behaviors appropriate for work innovation. Many different approaches can be used, from regular orientation sessions provided by management to training provided by previous participants. Training new members is a standard duty of semiautonomous work groups.

The parties should ensure that participants are provided with the requisite minimum of training in the new work behaviors before the work innovation is adopted. They should also create mechanisms to ensure that retraining is provided from time to time and that subsequent participants receive such training as well.

RECOMMENDATION ▷▷▷▷▷▷▷▷▷▷▷▷ 41

Gaining Commitment

Commitment to the new behaviors and their purposes creates strong motivation. When motivation is high, the behaviors are performed voluntarily. In one case it was found that when the behaviors were the result of voluntary, personal choice, a program of semiautonomous work groups thrived and grew. Later, when voluntarism lagged (for extraneous reasons) and management *required* others to participate, the program declined.[4]

Commitment is rarely the result of a single action, but rather the combined result of many: adequate training, for example, so that purposes and individual roles are clear to employees; the equitable distribution of rewards; and the achievement of congruence between the new and old.

Complete commitment of top management sets the stage for all that is to follow. Management has to invest time, resources, and personal attention. Weak signals, or dissension about the desirability of the innovation, will transmit itself down the line.

Other research has illuminated how commitment works toward institutionalization.[5] When there is strong belief in the new approaches, an attack on those beliefs will strengthen them. But if the beliefs are weak in the first place, similar attacks will weaken them further.

Again, as in the need to train newly assigned employees, turnover can enter the picture. Getting commitment in the beginning and holding turnover down will offset the weakening effect of later turnover. There will be less chance that the doubts of new employees will shake the motivation of the whole team, and they are more likely to be worked into the new system. So it is better to keep the entry of new employees at a minimum until later in the program, when it is firmly established.

Work innovations that provide the opportunity for *re*commitment, particularly through public gestures, have a better chance of surviving. Scanlon Plans, for instance, call for periodic referenda, and a vote of 90 percent or more may be set as a condition for continuation. Or there may be a public display of support for the innovation when a site welcomes a new top manager. Efforts by the parties to "sell" the value of the change to other organizations also amount to a *re*commitment.

RECOMMENDATION 42 ◁◁◁◁◁◁◁◁◁◁◁◁ *Top leadership of the employing organization (and the union) should demonstrate their commitment to the innovation, publicly and tangibly, in order to encourage voluntary participation and create staying power.*

Allocating Rewards

In many ways, the system of rewards determines the culture of the organization. If managers and other employees are urged to follow a QWL philosophy while rewards are issued for traditional behavior, there can be no doubt about which will prevail. Top management's commitment is demonstrated by what it does as well as by what it says. General Motors, for example, publicly praises and promotes managers who have combined outstanding economic performance with outstanding success in work innovation. General Foods, on the other hand, in effect isolated the Gaines pet food plant by ostracizing the managers who had successfully established its highly profitable experiment in job redesign.

Rewards are classified as intrinsic or extrinsic. Extrinsic rewards, such as promotions and pay increases, are given to the employees by someone else. Intrinsic rewards arise within the individual as a result of the improved work environment itself. The employee enjoys the sense of being trusted to exercise responsibility rather than serving as a mere adjunct to a machine; he or she may feel rewarded by feelings of accomplishment, freedom, or the opportunity to perform a variety of tasks.

Many organizational-change programs have been based on the assumption that intrinsic rewards are sufficient for institutionalization, but experience proves them wrong. Recent studies show that programs combining extrinsic and intrinsic rewards have attained the highest degree of institutionalization.[6] This finding is consistent with common sense and with national quality-of-work-life surveys, which show that employees want *both* kinds of rewards, not one or the other.

In order to have the desired effect, rewards must be clearly linked to actual performance of the changed behaviors rather than to merely going through the motions. This applies to managers as well as to other employees.

Another frequent block to institutionalization is employees' perception that rewards are unfairly distributed or withheld. (That perception is also a frequent source of collective bargaining problems.) The inequity may be a failure of the employer to make an expected payment, or it may be a lack of evenhandedness between one group of employees and another, or between managers and employees.

In one study of a job-enrichment program that was not institutionalized, the major reason for failure was that the workers were not compensated financially for the new skills they had learned.[7]

While they had not been promised more money, the fact that they were accomplishing more, but with the same pay, seemed inequitable to them.

In the case of Rushton Mining Company, where coal miners in an experimental section were organized into semiautonomous work teams, all members of the crew were paid at the same (but higher) rate, whereas before there had been sizable pay differentials between skills. Those who used to be paid more highly because they had longer experience, resented the elimination of this differential. The change program declined.[8]

RECOMMENDATION 43 ◁◁◁◁◁◁◁◁◁◁◁◁◁◁

The parties should include in the strategy of work innovation a system of rewards, both intrinsic and extrinsic, linked to actual performance of desired new behaviors and equitably distributed. However, actual payments of extrinsic rewards should never be made in advance of the achievements on which they are contingent.

Diffusing Change

The need to diffuse change arises in two different ways. Some work innovations, such as gain-sharing plans and certain socio-technical designs, are introduced throughout a site from the beginning. If a gain-sharing plan is designed to measure performance and provide rewards on a site-wide basis, it is impractical to install it piecemeal. Or, where managers as well as workers are to be organized in semiautonomous work teams, it may be advantageous to establish the entire structure at once and let it evolve as a whole. The problem of diffusion then arises only if the site in question is one unit of a multisite enterprise.

Most commonly, however, QWL projects, quality circles, and others, are pilot-tested in one or a few parts of a site until success is assured and other parts of the site want to emulate them.

It has generally been assumed that a change program will not become institutionalized unless it is diffused throughout an organization. Richard Walton's study of the Gaines pet food experience has persuaded him that a successful site-wide project can survive *without* diffusion to other sites in the enterprise, provided that corporate management does not actually suppress it. He qualifies this finding by pointing out that the successful model was reinforced by emulation in *other* enterprises.[9]

With respect to diffusion within a site, the pace of change is of paramount importance. If the change remains restricted to one or a few parts of the organization too long, it may, after a while, appear static or not be taken seriously. It may come to be viewed as an isolated practice rather than the desired mainstream way of doing things. On the other hand, if momentum develops, accompanied by a general perception that change is coming, interest will be maintained because it is going to affect everyone sooner or later.

Too slow a diffusion damaged the program at Rushton Mining Company. Having been confined to one section, it began to be perceived as a fluke and thus inappropriate for the rest of the organization. The same phenomenon has also been reported in several large manufacturing companies.

But there is also a risk in too rapid diffusion. Widespread understanding, adequate resources, and prior training must undergird expansion. Understanding and acceptance must be built ahead of time as a base from which to move to other areas of the organization. If there is excessive haste, the external forms of change, but not the real substance, are put in place and the structure collapses.

The right speed will vary with the organization, depending on the degree of change involved, the technology in use, and the climate among employees and managers. The rate of diffusion is not a final mathematical value. It will vary between plants and offices and between enterprises. However the speed requires clear and continuing attention as a factor in sustaining the right momentum. One should avoid both "cul de sacs" and headlong plunges over the precipice.

Early planning should encompass the diffusion of change as well as the design and implementation of the new program. If responsibility is assigned only for initial installation, there may be too great a lag before the organization is prepared for diffusion. The task is to change a whole organization, and to accomplish it requires a strategy.

Feedback, adjustment . . . feedback, adjustment . . .

The process of work innovation involves more than designing a program and putting it into operation. Actual operation may diverge widely from the original design. Management may not have what it thinks it has. Innovative programs may all go under the same name but may vary considerably when looked at up close. Some may prove to be changes in name only, and inspection will show that they have reverted to traditional practices.

On the other hand, there may be good reason to alter the initial design and make changes as the program unfolds. It may have to adjust to unforeseen realities of group dynamics and/or technology. To make these adjustments requires regular feedback of information.

"Sensing" and "correction" are the procedures by which the organization finds out how well the program is doing and takes steps to correct problems as they emerge. In organizations with the most complete institutionalization, there exist mechanisms for regular feedback of information, and adjustments are made on the basis of that information.[10] Change frequently does not become institutionalized because there is a lack of feedback and adjustment.[11]

In another case, adequate information was available, but no one acted upon it. Responsibilities must be assigned for gathering, communicating, and responding to information about the progress of change.

RECOMMENDATION 44 ◁◁◁◁◁◁◁◁◁◁◁◁◁

The parties should include in the original strategy mechanisms for creating a flow of information from the program and making adjustments as they are needed. Employee commitment and involvement are sustained by a free and steady flow of information.

Using Outside Information

There is a growing demand for consultants who help organizations install work-innovation programs. Quite a few have gained experience from seeing innovations work in different settings. They can be useful in getting up to speed on new developments, in identifying options, and avoiding pitfalls others have encountered. They also have experience in helping managements convey the new approaches to employees.

With regard to institutionalization, the problem is not in using consultants, but in using them too much or for too long. If too much of the program depends on their efforts, continuation of the program after they leave will be endangered. The greater the dependence on the consultant, the greater the risk to the program. In some cases, such as Harman International Industries in Bolivar, the consultants were retained too long because the parties did not really want the program to continue, but felt compelled to keep it alive.[12]

Consultants can be used to build in capability rather than provide an exclusive outside source for it. They can bring the outside source in and act as counselors to the internal managers who have been assigned the responsibility for design and implementation. They can help in the initial period of design and implementation and then be phased out.

The parties should use consultants when necessary but not let the program become too dependent on them. The consultant's highest contribution is to help create an internal capability, rather than become a substitute for it. Consultants should adapt to the organization rather than require the organization to be refashioned in the consultant's mold.

RECOMMENDATION ▷▷▷▷▷▷▷▷▷▷▷▷ **45**

The Indispensable Sponsor

As in the case of consultants, there can be overdependence on an individual in the organization, typically the executive or union leader who initiates the program and becomes its principal backer. On the one hand, such an individual is an advocate and the source of the power and resources that create and sustain the program in its early days. On the other hand, a program overly dependent on a sole sponsor becomes vulnerable if that sponsor leaves, especially in the formative years of change.

Changes in sponsors are among the most sensitive and significant decisions in protecting the lifeline of innovations. Frequently sponsors are transferred or promoted at the worst possible time in the phasing-in of the program, in order to satisfy career-development goals. This action is read as a lack of top management commitment, a strongly negative signal.

There are many examples of programs declining or being terminated because sponsors left. One case is that of a major organizational intervention in the State Department which persisted over a period of years but declined as soon as the initiator left. The new administrator was not sympathetic to the values and structure of the change program. A similar decline was noted when an innovative college president left after instituting a new structure for the school. There are two ways of avoiding this hazard.

The principal sponsor or sponsors, having originated the work innovation in the first place, should also take responsibility for embedding it in the organization. It is up to them to transmit their conviction to the entire organization, to make the change an essential part of the organization's style, and not to let the project become identified with one or a few individuals. Management should also make sure that deputies and back-up sponsors are developed to avoid loss of leadership or breaks in program direction when a sponsor must leave.

RECOMMENDATION
46 ◁◁◁◁◁◁◁◁◁◁◁◁◁◁

The parties should ensure that a work innovation does not become permanently identified with its sponsors. However, in the early days of a project, when the role of the sponsor may still be vital, the parties should transfer or move the sponsor only if they have a replacement who is willing and able to carry the project forward.

NOTES

1. Richard E. Walton, "The Topeka Work System: Optimistic Visions, Pessimistic Hypotheses, and Reality," in *The Innovative Organization: Productivity Programs in Action*, edited by Robert Zager and Michael P. Rosow, Pergamon Press/Work in America Institute Series (New York: Pergamon Press, 1982).

2. J.M. Ivancevich, "Changes in Performance in a Management by Objectives Program," *Administrative Science Quarterly* 17 (1972): 563-574.

3. Paul S. Goodman, *Assessing Organizational Change: The Rushton Quality of Work Experiment* (New York: Wiley-Interscience, 1979).

4. Paul S. Goodman and James W. Dean, Jr., "Institutionalization or Making Labor and Management Change Programs Last," an unpublished paper prepared for Work in America Institute, 1981.

5. C.A. Kiesler, *The Psychology of Commitment: Experiments Linking Behavior to Belief* (New York: Academic Press, 1971).

6. Richard E. Walton, "Establishing and Maintaining High Commitment Work Systems," in *The Organizational Life Cycle*, edited by J.R. Kimberly and R.H. Miles (San Francisco: Jossey-Bass, 1980).

Paul S. Goodman, *Assessing Organizational Change: The Rushton Quality of Work Experiment* (New York: Wiley-Interscience, 1979).

7. E.A. Locke, D. Sirota, and A.D. Wolfson, "An Experimental Case Study of the Successes and Failures of Job Enrichment in a Government Agency," *Journal of Applied Psychology* 61 (1976): 701-711.

8. Paul S. Goodman, "The Rushton Quality of Work Life Experiment: Lessons to Be Learned," in *The Innovative Organization: Productivity Programs in Action*, edited by Robert Zager and Michael P. Rosow, Pergamon Press/Work in America Institute Series (New York: Pergamon Press, 1982).

9. Richard E. Walton, "The Topeka Work System: Optimistic Visions, Pessimistic Hypotheses, and Reality," in *The Innovative Organization: Productivity Programs in Action*, edited by Robert Zager and Michael P. Rosow, Pergamon Press/Work in America Institute Series (New York: Pergamon Press, 1982).

10. Paul S. Goodman and James W. Dean, Jr., "Institutionalization or Making Labor and Management Change Programs Last," an unpublished paper prepared for Work in America Institute, 1981.

11. Richard E. Walton, "Establishing and Maintaining High Commitment Work Systems," in *The Organizational Life Cycle*, edited by J.R. Kimberly and R.H. Miles (San Francisco: Jossey-Bass, 1980).

12. Barry A. Macy, "The Bolivar Quality of Work Life Program: Success or Failure?" in *The Innovative Organization: Productivity Programs in Action*, edited by Robert Zager and Michael P. Rosow, Pergamon Press/Work in America Institute Series (New York: Pergamon Press, 1982).

9.
Issues on the Horizon

The deliberations of the national advisory committee for this policy study brought to light several major issues which, while not yet of immediate concern, appear likely to surface as the work-innovations movement becomes firmly established in the United States. They are problems of success, but they will require no less care and thought than those resulting from economic stress. The three of greatest moment are employment security, gain sharing, and the spread of work innovations from the private sector to other sectors of the economy.

EMPLOYMENT SECURITY

American public opinion has always held that there is something unreasonable about workers who resist productivity improvements that threaten their jobs. Why the public should think so defies understanding. If it were costless for a worker to find alternative employment of equivalent status in the same locale, one might well argue that resistance was antisocial but, in most cases, the human costs are very high. Why should workers be asked to sacrifice for the good of society while their managers are rewarded? An employer cannot expect employees to go all out for productivity until they have some reasonable assurance that their efforts will not be used against them.

Therefore, when unions enter into a joint QWL program, they insist on an agreement that no employee will lose employment or earnings as a result of the program.

But the problem goes far beyond the effects of productivity improvement *per se*, which can usually be met through attrition. Traditionally, blue-collar employees have been managed as if they were ballast on a balloon: load them on when the craft is full of lift; toss them overboard when the craft is sinking. Nonsupervisory white-collar employees, as a rule, escape harm until the blue-collar work force has been cut to the bone, although in firms which have only white-collar employees, nonsupervisory employees tend to be treated like blue-collars. Supervisors and managers are the last to feel the blow, but they feel almost as insecure as their subordinates when a cut is threatened. Yet employers complain that employees show little loyalty to the firm.

Part of the reason for this readiness to lay employees off when the market turns soft is that employers have not thought through the real costs of the traditional policy. As J.F. Lincoln, former CEO of the Lincoln Electric Company, wrote many years ago, the costs of severance, recruitment, and retraining, when added to the operational imbalances caused by layoffs and the subsequent delays in getting up to speed after the downturn has ended, far exceed any immediate savings of wages and benefits. Other companies which have added up the figures agree. Lincoln also charged that the traditional policy allowed employers to get away with slipshod business planning and employee selection.

Long before Japanese industrialists discovered the advantages of "lifetime employment," the Lincoln Electric Company was following a no-layoff policy, which it has made the basis of a highly successful, highly productive business. Except for one instance of special circumstances immediately after World War II, the company has *never* laid anyone off. For the past two decades the Lincoln guarantee has been in writing.

Most employees and unions would settle for a good deal less. All they ask is that no layoffs take place until the employer has tried every reasonable means of avoiding them, and that when a layoff does occur, all sectors of the work force share the burden equitably. The 1982 Ford and General Motors agreements with the UAW represent a step in this direction.

Japanese-style lifetime employment is a bilateral commitment, although only the employer's side is usually portrayed. The *quid pro quo* is the employee's unspoken agreement to remain with the firm for his full career ("his" because in Japan women are regarded as temporary employees). An employee who leaves the company voluntarily faces extreme difficulty in obtaining a new job with another lifetime-employment company. Few Americans would be prepared to make such a commitment.

Unions have been rethinking their approach to job security. They will continue to seek legislative protection against the effects of plant closures and removals, but they will give first priority to collective bargaining arrangements. In the past they have acquiesced in management's right to lay people off and have tried to mitigate the hardships by means of governmental unemployment insurance and employer-paid supplemental unemployment benefits. These measures provide partial income security for idleness. The unions' new goal is likely to be employment security for productive purposes. They realize that in the long run partial pay for idleness not only reduces the purchasing power of their earnings but weakens the employer and damages the self-respect of employees. They realize also that job security, in the sense of keeping a particular assignment, does not fit today's competitive environment. If individuals are assured of continuing in employment at an equivalent rate of pay, they become far more willing to grant the employer the increased flexibility of task assignment that may be required for productivity.

The key to maintaining job stability and ensuring continuous employment for all employees—including managers—is for the organization to adopt these objectives as part of its overall strategy. They cannot be attained as afterthoughts. The employer needs to be ready with measures to counter either a short-term or a long-term decline of business activity, and it would be wise to consult with the union for the purpose.

Among the strategies which Lincoln Electric Company has found helpful are the following:

☐ Holding prices low enough to ensure a steady long-term growth of sales.

☐ Extending the guarantee only to employees with at least two years' service (although it has never had recourse to this exception).

☐ Hiring new employees only when long-term growth prospects make it essential (which means not hiring to meet a temporary surge).

☐ Hiring only individuals who, after careful screening by senior managers, are expected to make good permanent employees.

A variety of other tactics for reducing the cost of continuous employment, both short term and long term, are outlined below.

Short-Term Measures

When an employer anticipates, or is unexpectedly confronted by, a dip in activities that is expected to be of short duration (about three to six months), it can take several measures other than layoffs. The simplest is work sharing, an agreement with employees (or union) to reduce work hours and pay proportionately.[1] This permits the burden to be distributed equitably across the work force, preferably including white-collar as well as blue-collar workers, instead of putting some employees on full layoff while the remainder are employed full time. Although many union contracts allow the work-sharing option, it is rarely invoked because senior employees feel they have "paid their dues" by accepting layoffs in the past and are therefore entitled to preference now.

Not long ago, United Air Lines and the flight attendants' union averted over 200 short-term layoffs by allowing pairs of attendants to share jobs. Each pair, often consisting of one long-service and one short-service employee, worked out its own division of hours and journeys but guaranteed the airline that all flights would be adequately covered.

Many more work forces would probably be willing to accept work sharing if unemployment insurance could be applied to make up some of the loss of income. The State of California instituted a short-term compensation law in 1978 to meet that requirement, and analysis of the first three years' experience indicates that employers, employees, and unions have used it to level out temporary valleys without significantly raising the costs to taxpayers. Arizona and Oregon recently adopted similar legislation; New York and other states are considering legislation; and Congress now has before it a bill to provide technical assistance to encourage states to experiment with short-term compensation.

Some observers have asked: If it makes sense to use unemployment insurance for the relief of employees who accept shorter working hours and spend their nonworking time on personal affairs, would it not make even more sense to use unemployment insurance dollars to have them spend the nonworking hours *at their place of employment* in ways that increase their economic value to society, to the employer, and to themselves? Those hours could be spent, for example, in additional training, in upgrading skills, in learning how to participate more effectively in consultative or decision-making groups. Such activities can, in theory, be conducted during periods of normal business demand but, in fact, they tend to be slighted. Even if they are performed during normal periods, there is no reason why they should not be continued at other times as well.

Other uses of temporarily redundant employees have been devised by major Japanese companies. Some assign their most expert workers to act as productivity consultants. With workshops, tools, instruments, and engineering advice made available, these workers have free rein to roam the plant, identify problems, and work out solutions. They develop new tools and machines. Occasionally they take additional instruction on their own time.

Many firms in the United States and abroad use slack periods to catch up on deferred maintenance work, ranging from machine overhauls to painting. During such times Lincoln Electric Company takes on an increased volume of orders for reconditioning used welding machines (its own and those of other companies) and also introduces new products.

Some companies use subcontracting as a deliberate stabilizer, issuing contracts when business surges and taking the work back in-house when business declines.

A particularly ingenious practice is that of exchanging employees temporarily between plants or firms of different industries. German and Japanese employers, in particular, have used this device. A firm whose business is temporarily slack finds a nearby firm whose volume is high and for whose work the unneeded employees can be trained fairly quickly. If the loaned jobs pay less than the original one, the permanent employer guarantees the earnings of the transferees. While such loans have usually taken place on a bilateral basis, it is conceivable that a permanent multilateral scheme, based on the insurance principle, might work even better.

Long-Term Measures

Coping with a six-month or longer period of downturn, whether due to one industry's cycle or to a general economic recession, may require other measures. However, there seems to be no reason in principle why unemployment insurance should not be paid to redundant, work-sharing employees who are receiving training from their employers for a long period as well as for a short one, particularly when a recession has been deliberately engineered by the federal government. To prevent the waste of funds, the government might have to set a condition that the employer demonstrate that the business is not in permanent decline.

If the organization believes that its work-force requirements will not eventually return to "normal," it may consider strategic moves into other products or markets as a means of keeping employees gainfully occupied. Some Japanese employers have gone this route and Control Data Corporation is considering it, on the theory that it is worth suffering a loss on a new enterprise for a year or two in order to maintain the commitment of continuous employment. Since efforts to put together a new venture hastily can lead to disaster, such a strategy clearly requires long advance planning. As noted previously, however, American employers are much less willing to gamble money on investments to stabilize the work force than on other kinds of investments.

Employers should strengthen employment security in order to create the basis for employee confidence in, and cooperation with, work innovations. Their policies should balance the need for flexibility with the need for security, **RECOMMENDATION** ▷▷▷▷▷▷▷▷▷▷▷▷ **47** *recognizing the investment in human capital represented by training, experience, and established relationships. Where there is a union, these trade-offs should be collectively bargained.*

GAIN SHARING

Certain work innovations have a built-in gain-sharing procedure, based on the principle that monetary rewards should reinforce desired behaviors. Thus Scanlon and similar plans reward teamwork and group creativity. The system of pay-for-knowledge rewards an individual's efforts to become a more versatile member of a self-managing team.

The majority of consultative programs, whether or not they make use of groups, contain no explicit gain-sharing feature. Some programs may have a general understanding that if gains accrue, they will be shared. Others rely on intrinsic rewards alone. Still others, having been introduced when the employer faced an economic crisis, finesse the question of gain sharing because the prime concern is to keep the firm afloat and save jobs, without regard to extra earnings. The strength of job security and intrinsic rewards often suffice to keep a QWL program going for a long time. In some cases, employees may consider the improvement of relations with the supervisor a priceless gain in itself.

Sooner or later, however, when it becomes apparent that work innovations have been instrumental in lifting the employer out of crisis and into profit, employees are bound to ask: Are we getting our fair share? They will make it clear that gain sharing means extra rewards in addition to, not in substitution for, satisfactory basic wages and benefits.

At that point the organization must decide what its objectives are. Does it wish to reward everyone for organization-wide improvements? Or does it wish to reward individuals for outstanding contributions? Or both? The two are not mutually exclusive; each has its own virtues and defects. Organization-wide rewards for productivity improvement convey a sense of equity and foster cooperation, but they may stifle individual ambition, risk taking, and extra effort. Why should I do more than others if we all share alike? Individual incentives, on the other hand, although they spur extra effort and ingenuity, may pit one employee against the others, incite behavior that impairs the quality of working life, or even harm the productivity of the organization. Lincoln Electric is one of the few employers known to offer both kinds of gain sharing, with the result that average earnings are substantially above those of any other employer of comparable economic situation in the area.

Multisite employers have the further problem of choosing between corporate and site gain-sharing plans. Corporate profit sharing has the advantage of not requiring the books to be opened any wider than reporting to shareholders now requires them to be. The disadvantages are that (1) profits do not necessarily reflect productivity improvement because so many extraneous factors (such as accounting practices, tax peculiarities, and currency translations) intervene; (2) corporate sharing rewards low-productivity sites to the same extent as high-productivity sites, which may negate the objective.

Site-wide plans do require the books to be opened wider, as mentioned above, but that is more a psychological than a real hazard. Multisite employers, such as Dana Corporation, Midland Ross, and Parker Pen, have been able to live quite comfortably with disclosure for the purposes of Scanlon Plans. Their experience also shows that each site can work out its own sharing without conflicting with other sites.

When considering site-wide plans, the employer will want arrangements that treat managers and nonmanagers according to the same formula. The thrust of work innovations is to eliminate privileges based on rank.

Some quality-circle and similar programs reward groups or individuals for special contributions by processing their suggestions through the regular suggestion system. This procedure leaves evaluation up to an arm's-length group of managers. It sets the suggestors and the evaluators apart, as if the suggestors were trying to sell a finished product, whereas the relationship ought to be one of mutual assistance in developing an idea for the organization. Moreover, the idea itself may have been a joint product of management and employees, which implies that rewards should be shared among all who had a hand in it.

The effectiveness of a gain-sharing plan depends on whether the rewards are of a kind and amount that will elicit the desired response from employees. For example, at Harman International Industries in Bolivar, Tennessee, workers preferred to take their productivity bonus in the form of leisure time rather than money.[2] At Rushton Mining, the preferred form was dental insurance coverage.[3] At Shell Canada's Sarnia refinery, workers chose the opportunity to acquire additional skills as the route to pay increments.[4] In developing a plan, therefore, the employer should incorporate the views of the work force. Where there is a union, collective bargaining is essential; where there is not, it may be necessary to conduct attitude surveys or to have supervisors sound out their subordinates.

Employers engaged in work innovations should be sensitive to the long-term gain-sharing issues which **RECOMMENDATION** *are bound to arise. In cost-reduction* ▷▷▷▷▷▷▷▷▷▷▷ **48** *sharing plans (with or without unions), a formula is defined and agreed on beforehand. In other programs, gain sharing should remain a secondary goal, but as significant savings and improvements are generated, the employer should agree with the employees (through their union, if there is one) on a method for periodic sharing of financial gains directly attributable to the work innovations.*

THE ADOPTION OF WORK INNOVATIONS IN OTHER SECTORS OF THE ECONOMY

Since the productivity crunch is most apparent in the private sector of the economy, it is here that most of the effective work-innovation programs have been concentrated. This sector has been motivated to undertake and to stay with innovations primarily by the profit motive, which establishes a clear relationship between corporate performance and rewards and penalties.

In fact, the private sector is being measured constantly, with the consumer as critical arbiter of the product or service. Customer approval, translated into profitability, results in dividends to shareholders, healthy growth rates, return on investment, and rising common share prices. It also results in such managerial perquisites as executive bonuses, stock options, and deferred compensation. Customer disapproval, on the other hand, has resulted in the bankruptcy of thousands of small firms and even giant multinationals.

It is no wonder, then, that the search for productivity in the private sector has constituted a powerful incentive for the introduction of work innovations. Such incentives are largely absent, however, in most of the health-care sector, in education, and in government—despite the fact that adaptations of work innovations that have succeeded in the private sector could be an effective means of raising productivity.

The most important reason for using work innovations in these three sectors is because their work organizations are all highly labor intensive. They employ millions of people and consume a sizable part of the gross national product. The payrolls of these employers may represent as much as 70 to 80 percent of their budgets. It is particularly in such labor-intensive situations that the greatest opportunity exists to raise productivity through the more effective use of human resources.

It is noteworthy that in two of the three sectors, professional employees who are central to the effectiveness of these institutions—nurses and teachers—have been deprived of a voice in their management and are openly disaffected. Addressing the legitimate needs of these two groups through the adoption of innovative work practices would also serve the productivity goals of the enterprise.

Health Care

The provision of health care has become a major growth industry, employing approximately 3 million people. As health costs soar (federal outlays for national health needs provided directly to individuals are expected to reach $80 billion by 1984), the industry faces pressures from steadily rising consumer demands, high inflation, and the cost-containment efforts of government and major employers. These pressures are expected to persist through the coming decade. Hospitals are urgently seeking ways to meet these demands with greater effectiveness and efficiency—in short, with greater productivity.

Nurses are central to the productive operation of the hospital, but their contribution has long been taken for granted. In a well-run institution, they should serve as the link between the different components of patient care. More often than not, however, their responsibilities far exceed their authority: they are subject to the political power of doctors, the managerial direction of administrators, and the growing influence of blue-collar unions. The resultant sense of powerlessness breeds depression, which is compounded by the hospitals' seeming indifference to nurses' professional and personal needs.

Not surprisingly, the strong motivation that first impelled nurses to enter the profession no longer holds them there. High absenteeism and turnover, low morale and, ultimately, withdrawal from hospital nursing are the consequence. As word gets around, recruitment into the profession dries up.

A recent article on QWL in Canadian hospitals highlights some of the problems and possible solutions:

> A 1980 manpower study conducted by the Alberta Hospital Association identified factors causing nurses to leave. They include the inability to provide safe patient care, unsatisfactory shifts, and the lack of opportunities for continuing education and advancement. Nurses who have left the profession felt that they were forced to choose between their family and their jobs; work and family demands were just too great to handle.

> Working conditions, which include patient load, stress, and hours, are the reasons why 32 percent of the inactive respondents stated they will never return to nursing. Two-thirds indicated that they would return to nursing if certain conditions, such as availability of part-time work, choice of shifts, and fewer hours or more flexible hours, were fulfilled. Another condition was more time for patient contact and involvement in patient-care decisions. The study notes, "Unless nurses receive the satisfaction from being able to provide high quality patient care which attracted many of them in the first place, they will not return."[5]

In late 1981 the National Commission on Nursing, an independent body of doctors, nurses, administrators, and others, issued its initial recommendations to improve the status and morale of the nursing profession. Under the rubric Nurses and Health Care Institutions, the Commission stated, among other things, that nurses in practice should work collaboratively with other health-care professionals in care-delivery systems to:

□ Participate in developing organizational structures that provide nursing involvement in policy decisions and patient-care decisions related to nursing.

□ Determine organizational mechanisms that support professional practice, improved communication with other health professionals, and innovation to improve patient care.

□ Participate in setting standards for support systems for non-nursing patient-care-related functions.

□ Promote involvement in personnel policies, staffing and scheduling policies, and wage and benefit structures to meet needs of practicing nurses.[6]

Further, it advocated that "models of organizational design for nursing management units that provide mechanisms for nurse participation in clinical and managerial decisions need to be developed and/or evaluated for their impact on quality of care and nurse satisfaction."[7]

The solutions advocated by both sources are a call for work innovations in the health-care sector. Hospitals are not the same as the shop floor or the office, but the aspirations and needs of the "players" have much in common. These statements give good reason for hope. Grass-roots efforts to implement them are now being organized and will bear watching.

Education

The general perception of a large portion of Americans is that just when they are putting more resources into the schools, they seem to be getting less and less out. The perception is fed regularly by news coverage, often casually read, sometimes casually written. There is, however, enough to be concerned about in measured trends and competently gathered statistics.

□ *Educational achievement.* Scores on the Scholastic Aptitude Tests have been declining for over 15 years. The National Assessment of Educational Progress reports significant declines in mathematics and science among 17-year-olds, although achievement is holding up in reading and writing.[8]

☐ *Dropout rates.* The graduation rate of 18-year-olds peaked at 76 percent in 1967 and declined to 74 percent in 1980.[9]

☐ *Attendance.* Although, on a nationwide average, 92 of every 100 students enrolled showed up in classrooms in 1978-79, the average for the 20 largest cities was only 87 out of 100, with lows of 70 per 100 in Boston and New York.[10]

☐ *Discipline.* Twenty-six percent of Americans in 1977 regarded "lack of discipline" as their prime concern about public education. One teacher in eight claimed to have suffered physical attack or property damage.[11]

☐ *Confidence.* In 1979, only 28 percent of Gallup Poll interviewees registered "a great deal" of confidence in those who ran the educational system, as compared to 49 percent in 1974 and 37 percent in 1973.[12] Only 18 percent of parents of public school children thought the schools were doing an excellent job.[13]

Although all these problems involve the classroom, teachers nowadays rarely receive more than a passing mention in treatises on educational improvement. They have come to be viewed as a purely passive part of the teaching machinery, overshadowed by considerations of textbooks, school facilities decisions, research pointing to the primacy of "background factors," state-wide minimal competency testing, augmentation of guards in high school hallways, and drug bust stories.

Large schools, with up to 4,000 students and large bureaucracies to manage "teacher input," have made individual initiative harder to nurture and caused alienation to grow. Relations between students and teachers, teachers and administrators, parents and teachers, and parents and administrators, are marked by conflict.

In May of 1979, the New York State United Teachers surveyed its membership to identify the causes of stress among teachers. The most stress-producing factors were "managing 'disruptive' children" and "incompetent administrators." Next on the list was "maintaining self-control when angry." Urban teachers reported three times as many stressful items as rural teachers, and twice as many as suburban teachers. Teachers 31 to 40 years old seemed to be under the greatest stress, an age at which many are giving up teaching for something else.[14]

A poll by the National Education Association in 1980 found that one-third of the teachers sampled were dissatisfied with their jobs, and 41 percent would not become teachers if they could start their careers again.[15] By contrast, fewer than 12 percent of the U.S. work force as a whole say they are dissatisfied with their jobs.[16]

So much alienation in any professional group is painful; but when it afflicts those on whom we rely to educate our youth, it is clear that corrective action is needed quickly.

The conditions of schools and learning, and the narrowing of teacher discretion and opportunity for initiative, reveal a pattern not too unlike emerging trends in business organizations. It is not surprising that these similarities exist, if only because both operate through large organizations and bureaucracies.

Given the complexity of relations within the school and of relations between school and community it would be imprudent to apply industrial experience directly to education. Nevertheless, work innovations clearly can play a useful part in school operation. The occupational role of teachers—especially secondary school teachers, who have the most stressful jobs—offers a strategic point of entry. Already school systems in New York State and Michigan have made good use of labor-management committees, and a few schools have tested quality circles for teachers, but this is barely a beginning.

Government

Federal, state, and local governments combined employ roughly 18 percent of the work force. At first glance they appear to be tempting targets for improvement of productivity through large-scale application of work innovations, but a more careful look suggests caution. Governments carry out an enormous diversity of activities, many having inherently irreconcilable objectives, so that concepts of productivity are irrelevant. In many areas the notion of service to the public is altogether lacking: often the "customer" is the legislature or the executive—whichever one happens to be the immediate source of funds. Across-the-board programs to increase productivity in government, therefore, tend to be cost-cutting exercises rather than efforts to give the public more value for its money.

On the other hand, some government activities do have well-defined end-products or services, delivered to well-defined bodies of recipients, with reasonably well-defined costs. In principle, these should be prime candidates for productivity improvement, since reasonably objective assessments of their performance can be made.

Any effort to raise productivity in such agencies, however, runs into severe obstacles, over and above the absence of a profit motive:

□ It is hard to use gain sharing as a motivator in the public sector.

□ Communications, especially in the federal government, are impeded by political sensitivity and the classification of documents.

□ By tradition, organizational size counts more than effectiveness as the basis of rewards.

□ Managers believe that any attempt to deploy people more productively will be baffled by civil service regulations and/or union contracts.

□ The high turnover of managers concentrates attention on short-term actions, at the expense of the long term.

□ Since few leading figures in the public sector have had previous experience of management, they have difficulty in understanding how an organization works or how work innovations might help.

□ Many managers share the belief that civil servants, by and large, are time servers, indifferent to the consequences of their work. Work innovations succeed only where management is willing to act on the opposite assumption.

We are aware that there have been remarkable programs in the public sector, such as the federal program of alternative work schedules; quality circles at the Norfolk (Virginia) Navy Yard; the New York State Employees' joint quality-of-work-life program; the California short-term compensation (work sharing) law; management-union QWL programs in several Ohio and Massachusetts cities; and area labor-management committees in Jamestown, New York, and a dozen other centers. None of these programs, however, have attained the recognition accorded private-sector work innovations such as those at General Motors or Topeka—the kind of recognition that impels others to emulate them. The few pockets of change almost vanish in the vast universe of opportunity and need.

Under the circumstances, one should perhaps take heart from the fact that any successful work innovations have been introduced in government at all, rather than be discouraged by the small number. But the matter cannot be left there. As the private sector—for profit and not for profit—increasingly devotes its resources to productivity improvement, the public will clamor more and more loudly for governments and their millions of employees to do likewise. Rational solutions for the built-in obstacles will have to be found. The tasks ahead, then, will be to develop distinctively public models of success, to gain credibility, and to disseminate guidelines for establishing these work innovations in the public sector.

The three issues reviewed above will figure largely in the development of the work-innovation movement over the next decade.

Employment security has long been the dominant inducement for workers to participate in decision making. Since persistent recession and unemployment will remain vivid in memory for years to come, security will continue to be an issue.

Gain sharing will gain in importance, not only because Scanlon and similar plans are useful points of entry, but because other forms of work innovation will bring financial gains to employers and workers will demand equity.

Lastly, there will be a growing demand for work innovations in the public and not-for-profit sectors, matched by the growing force of example from the for-profit sector.

How rapidly these become live issues in the workplace will depend partly on the direction of the economy and partly on the rate at which work innovations "take" in all sectors. The remainder of the eighties will tell the tale.

NOTES

1. Work in America Institute, *New Work Schedules for a Changing Society* (New York: Work in America Institute, 1981), pp. 88-105.

2. Barry A. Macy, "The Bolivar Quality of Work Life Program: Success or Failure?" in *The Innovative Organization: Productivity Programs in Action*, edited by Robert Zager and Michael P. Rosow, Pergamon Press/Work in America Institute Series (New York: Pergamon Press, 1982).

3. Richard E. Walton, "The Rushton Quality of Work Life Experiment: Lessons to be Learned," edited by Robert Zager and Michael P. Rosow, Pergamon Press/Work in America Institute Series (New York: Pergamon Press, 1982).

4. Stanley D. Nollen, *New Work Schedules in Practice: Managing Time in a Changing Society*, Van Nostrand Reinhold/Work in America Institute Series (New York: Van Nostrand Reinhold, 1981), pp. 83-93.

5. National Commission on Nursing, *Initial Report and Pre-*

liminary Recommendations (Chicago: National Commission on Nursing, 1981).

6. Lucie Brunet, "Quality of Working Life in Hospitals," *Quality of Working Life: The Canadian Scene*, January 1981, pp. 12-16.

7. Ibid.

8. The College Board, *On Further Examination*, Report of the Advisory Panel on the Scholastic Aptitude Test Score Decline (Princeton, N.J.: The College Board, 1978).

9. National Center for Educational Statistics, *The Condition of Education* (Washington, D.C.: National Center for Educational Statistics, 1980), p. 80.

10. Ibid., p. 82.

11. Ibid., p. 52.

12. National Opinion Research Center, *General Social Surveys, 1972-1978*, Cumulative Codebook (Chicago: National Opinion Research Center, University of Chicago, 1978).

13. "The Gallup Survey of Public Attitudes toward the Public Schools," *Phi Delta Kappan*, December 1975.

14. New York State United Teachers Research and Educational Services, *Information Bulletin, 1979* (rev. February 1980)

15. "Poll Shows Teachers Dissatisfied," *New York Times*, July 15, 1980.

16. Robert P. Quinn and Graham L. Staines, *The 1977 Quality of Work Survey* (Ann Arbor, Mich.: University of Michigan, Institute for Social Research, 1979).